So Simple a Beginning

So Simple a Beginning

HOW FOUR PHYSICAL PRINCIPLES SHAPE OUR LIVING WORLD

Raghuveer Parthasarathy

PRINCETON UNIVERSITY PRESS
Princeton and Oxford

Published by Princeton University Press
41 William Street, Princeton, New Jersey 08540
6 Oxford Street, Woodstock, Oxfordshire OX20 1TR

press.princeton.edu

All Rights Reserved

Library of Congress Cataloging-in-Publication Data

Names: Parthasarathy, Raghuveer, 1976– author.
Title: So simple a beginning : how four physical principles shape our living world / Raghuveer Parthasarathy.
Description: Princeton : Princeton University Press, [2022] | Includes bibliographical references and index.
Identifiers: LCCN 2021026503 (print) | LCCN 2021026504 (ebook) | ISBN 9780691200408 (hardback) | ISBN 9780691231617 (ebook)
Subjects: LCSH: Biophysics. | BISAC: SCIENCE / Life Sciences / Biophysics | SCIENCE / Life Sciences / Anatomy & Physiology (see also Life Sciences / Human Anatomy & Physiology)
Classification: LCC QH505 .P37 2022 (print) | LCC QH505 (ebook) | DDC 571.4—dc23
LC record available at https://lccn.loc.gov/2021026503
LC ebook record available at https://lccn.loc.gov/2021026504

British Library Cataloging-in-Publication Data is available

Editorial: Jessica Yao, Ingrid Gnerlich, and Maria Garcia
Production Editorial: Natalie Baan
Text Design: Carmina Alvarez
Jacket Design: Jessica Massabrook
Production: Jacquie Poirier
Publicity: Kate Farquhar-Thomson and Sara Henning-Stout
Copyeditor: Jennifer McClain

This book has been composed in Charis

Printed on acid-free paper. ∞

Printed in the United States of America

1 3 5 7 9 10 8 6 4 2

In memory of my parents,
Kalyani and Sampath Parthasarathy

Contents

Part III: Organisms by Design

So Simple a Beginning

Introduction

How does life work? This question may seem overwhelming, or even preposterous. How could any answer do justice to both a sprinting cheetah and a stationary tree, to the unique *you* along with the trillions of bacteria that live inside you? The experiences of even a single organism are breathtakingly varied: consider a chick's emergence from its egg, the first flap of its wings, the racing of its heart at the sight of a fox, and its transformation of food and water into eggs of its own. Could any intellectual framework encompass all of this?

The search for an answer—for some kind of unity amid the diversity of life—is reflected in our ancient urge to categorize living things based on similarities of appearance or behavior. Aristotle partitioned animals into groups using attributes such as laying eggs or bearing live young. Ancient Indian texts applied a variety of classifiers, including, similarly, manner of origin: "those born from an egg, those born from an embryonic sac, those born from moisture, and those born from sprout." Modern taxonomy emerged from the eighteenth-century work of Carl Linnaeus, who systematized the naming of organisms and developed a hierarchical classification scheme based on shared characteristics that we continue to find useful. Classification in itself, however, is not very satisfying. We want to know the *why*, not just the *what*, of the commonalities that unify living things.

In this book, we look for that *why* through the lens of physics, revealing a surprising elegance and order in biology. Of course, this isn't the only perspective that offers deep insights into life. There is the viewpoint of biochemistry, with which we understand how atoms join together to form the molecular components of organic matter, how energy is deposited in and extracted from chemical bonds, and how the incessant flux of matter and energy in chemical reactions constitutes the metabolism of living things. But it is difficult to use chemistry alone to zoom out from the scale of molecules to the scales of the animals and plants around us, or even the scale of single cells, and make sense of shape and form.

Another all-encompassing perspective is that of evolution. Since the mid-nineteenth-century epiphanies of Charles Darwin and Alfred Russel Wallace, we can see the traits of living creatures as manifestations of deeper historical processes. Similarities, whether of visible characteristics of anatomy or more hidden patterns in DNA sequences, can reflect shared ancestry with which we can deduce a tree of relationships linking all of life together. Differences emerge due to random chance and the varied pressures on survival imposed by creatures' environments; again, present forms reflect past history. Evolution provides a powerful framework for understanding life. It is not, however, one that we focus on in this book. In part, this is because there is already a large popular literature on the subject. More importantly, however, evolutionary principles alone don't illuminate the *why* as much as the *how*.

To illustrate what I mean by "why," consider the swim bladder, a pair of gas-filled sacs possessed by many, but not all, species of fish. Comparing creatures both extant and extinct reveals this organ's evolutionary history, with connections to the emergence of lungs in air-breathing animals that Darwin himself remarked upon. Understanding the function of a swim bladder, however, requires a bit of physics: the low density of the enclosed gas offsets the high density of bone in bony fishes, allowing the animal to maintain the same average density as its watery surroundings and thereby easily position itself at whatever depth it likes. A swim bladder is just one solution to the challenge of matching density. The fish might instead contain large amounts of low-density oil, or a skeleton composed of cartilage rather than bone, both of which are strategies adopted by sharks, which lack a swim bladder. The last common ancestor of cartilaginous and bony fish lived over 400 million years ago. Since then, the distinct evolutionary paths of the two groups have led to different solutions to the shared physical challenges of aquatic locomotion. We can state, with a point of view echoed throughout this book, that understanding the why of these anatomical features, related to control of density, highlights a hidden unity that fish share that transcends their evolutionary divergence. We should keep in mind, however, that the machinery of variation and natural selection—the enhanced odds of survival that accrued over generations to those creatures better able to navigate their aquatic world—provides the paths by which the forms we see arise.

There are other vantage points besides those of biochemistry and evolution from which to survey the breadth of life. Rather than list all the approaches we won't be exploring, however, let's turn to the one we will.

I've already hinted at the view of nature the rest of this book expands upon, which I identify as *biophysical*. The term implies a unification of biology and physics. It encapsulates the notion that the substances, shapes, and actions that constitute life are governed and constrained by the universal laws of physics, and that illuminating the connections between physical rules and biological manifestations reveals a framework upon which the dazzling variety of life is built. The notion of universality is central to the utility of physics, and to its appeal. The same principles of gravity apply to an apple falling from a tree and to planets orbiting the sun, and current work aims to further expand this framework to encompass the strange behavior of the quantum world. Biophysics extends to the living world the quest for unity that lies at the heart of physics.

To say that living things obey the laws of physics may seem trivial. After all, organisms are made up of the same fundamental particles that make up everything else and are therefore governed by the same rules. But one might expect the explicit role of physics to be over after physical forces set up the formation of atoms and molecules, with complex chemistry giving shape to further molecular rearrangements and the idiosyncratic predilections of cells and organisms being responsible for larger features. This is, however, incorrect. Just as physical forces direct the intricate branching of frost on a winter window and the rhythmic curves of vast desert dunes, and do so in ways that don't require subatomic particles for their explanation, physical mechanisms shape life at all scales. One of the great triumphs of physics, especially over the last half century, has been an understanding of how broad rules arise in all sorts of natural phenomena, clearing the underbrush of complexity to reveal deep principles. Magnets, for example, become nonmagnetic if heated above some specific "critical" temperature; though magnets can be made of many different elements and alloys, each with their own unique atomic-scale structure, the magnetic field of every magnet decays with exactly the same form as it approaches its critical temperature. Being a three-dimensional arrangement of inter- acting atoms, it turns out, suffices to determine the consequences of these interactions, regardless of atomic details. As another example, consider a shaken container of mixed nuts. One typically finds that the larger nuts rise to the top, giving this well-known phenomenon its name: the Brazil nut effect. The effect isn't particular to nuts, of course, and occurs in mixtures of cereal grains, rocks on riverbeds, and any collec- tion of agitated, disordered objects. Its explanation involves general no- tions of what are called granular flows, and the ways in which any ensemble of colliding particles must create and fill in interstitial spaces in order to move.

Biophysics applies this quest for broadly applicable physical rules to the world of living things. This endeavor, though still incomplete, has already been far more successful than we might have dreamed even a few decades ago. Using physics, we can understand the bursting of DNA from viruses, fundamental limits on the speed of thought, and the reg- ular spacing of our vertebrae. We can apply our insights to grow or- gans on slabs of plastic and read genomes using pulses of light. We un- cover a simplicity and an elegance in the living world that is otherwise hidden. Simplicity emerges because a handful of principles rather than a morass of detail suffices for many explanations; elegance because of the unity shared by the living and nonliving world. This is an unusual point of view; I hope the pages to come will convince you of it.

Every quest for unity amid complexity risks the pitfall of hubris, how- ever. There is the temptation to ignore the lessons that variety pro- vides, or to force motley data into unreasonably simple frameworks. A physical perspective is especially prone to these missteps, perhaps because of the elegance of its theories and perhaps because of their historical successes. Despite being a physicist myself, I note that the caricature of physicists as blithely trampling, elephant-like, through adjacent fields of inquiry without adequately appreciating the treasures underfoot is not wholly inaccurate. Though this book is a celebration of biophysics, I'll describe some of its stumbles as well; chapter 12 in particular examines contentious issues of metabolism against which a biophysical approach may have failed.

• • •

What are the physical principles that govern living things? We could refer to laws related to fundamental forces, thermodynamics, probability, and so on, amenable to precise mathematical formulation. While rigorous, this would be rather dry, and would moreover obscure the overarching lessons that biophysicists have drawn from nature. Instead, I direct our attention to four concepts or motifs that arise repeatedly in biophysical explorations.

The first is *self-assembly*, the idea that the instructions for building with biological components—whether molecules, cells, or tissues—are encoded in the physical characteristics of the components themselves. It may seem obvious that an organism contains its own instructions. After all, one doesn't need to carve a tree into a tree shape or paste five arms on a starfish; the creatures organize their own forms. Their internal instructions, however, need not take the form of a task list written into one set of components and executed by another. Rather, the physical characteristics of biological materials often *are* the instructions. Features like size and shape can guide the arrangement of pieces into a larger whole, as can less visible attributes such as electrical charge, harnessing the laws of physics.

I'll illustrate with an example. If you've ever blown soap bubbles and watched them come together, you may have noticed that there's never a junction at which more than three bubbles meet. Four adjoining bubbles may look like the drawing on the left of the figure below (page 6), with boundaries like a bent letter H, but never like the drawing on the right, with boundaries like an X. Physical forces drive soap films to minimize their surface area, leading to incontrovertible rules for

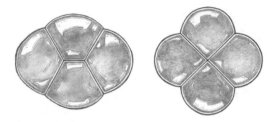

sets of bubbles that have been appreciated since their nineteenth-
century elaboration by Belgian physicist Joseph Plateau. These rules
prohibit any junction of four bubbles, as such a junction could never
be part of a minimal-area surface. The arrangements of bubbles aren't
haphazard. No external hand, however, is needed to guide them into
their stereotyped pattern; the rules for their organization are embedded in their physical nature. For well over a century, scientists
have noticed that arrangements of adjoining cells in all sorts of tissues
resemble the arrangements of soap bubbles, and have investigated
whether this is coincidence or a reflection of similar underlying mechanisms. In 2004, for example, Takashi Hayashi at the University of
Tokyo and Richard Carthew at Northwestern University looked at the
cluster of photoreceptor cells situated in each of a fruit fly's compound
eyes. Normally, there are four, with exactly the same arrangement as
four soap bubbles. Using mutant flies that developed 1, 2, 3, 5, and 6
photoreceptor cells per group, they found the same arrangements that
one finds in assemblies of 1, 2, 3, 5, and 6 adjoining soap bubbles. The
fly, it seems, relies on general physical mechanisms of surface area
minimization to organize these crucial cells of its retina. Rather than
painstakingly positioning cells, the fly makes the cells and lets them
sort out their contacts, minimize their areas, and pattern themselves
on their own. The cells, like the soap bubbles, assemble themselves.
In countless other contexts as well, we similarly find that structure
isn't drawn explicitly into the blueprints of an organism; rather, nature places the raw materials at the site and trusts that the laws of
physics will put them together properly. Thankfully, the laws of
physics are reliable workers.

The second recurring motif is that of a *regulatory circuit*. The ubiquity of computers makes us familiar with the idea that machines can use rules of logic to transform inputs into outputs, making decisions based on signals from sensors or controllers. We're also comfortable with the idea that living creatures, ourselves included, make behavioral choices based on the stimuli in their environment, though the details of the computations are more mysterious. We'll see that decision-making circuitry is not just a feature of the large-scale world but is manifested in the microscopic activities of life's molecules, built in to their very structure and modes of interaction. The wet, squishy building blocks of life assemble into machines that can sense their environment, perform calculations, and make logical decisions.

A migrating cell in a developing embryo, for example, must stop its wandering when it reaches the appropriate destination, a decision determined in part by assessing the mechanical stiffness of the neighboring tissue. Cells adhere using proteins that jut out from their surfaces, and through these proteins they can tug on their surroundings. Some adhesion proteins can serve as sensors as well as anchors, with these two roles inexorably linked: for stiff surroundings, the protein molecules are stretched, as your arm would be if tugging on a thick tree branch from a few feet away; for soft surroundings, the proteins are bent, as your arm would be if pulling a towel on a clothesline, easily dragged toward you. The cell contains other components that can bind to sites on the adhesion protein only if those sites are exposed, which occurs only if the molecule is stretched—imagine the inside of your elbow, accessible as you tug on the tree but not the towel. This binding triggers events that culminate in the cell's decision to stop its wandering. The physical conformation of the protein, therefore, underpins a cell-scale machine that senses, calculates, and decides.

Our third concept is that of *predictable randomness*. The physical processes underlying the machinery of life are fundamentally random but, paradoxically, their average outcomes are reliably predictable. In the nonliving world, randomness is central to activities as diverse as the shuffling of cards and the collisions of gas molecules. Physics has long tackled the question of how robust features emerge from underlying chaos. We know, for example, why steady, consistently colored light shines from stars despite their churning interiors, and how energy can be extracted from the violent combustion of gasoline. The microscopic world is subject to incessant, vigorous, and fundamentally random motion that DNA and other cellular components must deal with, and even exploit. We can deduce the probable outcomes of random processes, which in many cases provide simple explanations of superficially complex phenomena. A virus reaching a cell that it may infect, for example, doesn't need to think (even if it were capable of thought) about how to find the specific surface proteins to which it can bind; it is buffeted by random forces that drag it everywhere, ensuring that its chaotic trajectory will intersect its target. Your immune system also makes use of randomness, generating an enormous variety of receptor proteins that might, by chance, recognize invaders that have never before been encountered. We devote all of chapter 6 to the randomness of microscopic motion, which finds echoes in discussions of genes and traits where randomness is also built into the way life works.

Our final recurring biophysical motif is that of *scaling*, the idea that physical forces depend on size and shape in ways that determine the forms accessible to living, growing, and evolving organisms. That size, shape, and physics are related is well appreciated for artificial structures. It's hard to build big buildings, for example. Before the advent of steel frames and other modern inventions, to attempt great heights or large interior spaces was to tempt collapse, as the weight of a structure could overwhelm the support its walls could provide. Simply scaling up a small building, maintaining the proportionality of its dimensions, fails. In modern language that we elaborate in chapter 10, gravity and other forces *scale* with size in different ways that we need to account for when designing buildings. Scaling concepts are similarly reflected in the sizes and shapes of animals but extend to much more than mechanical concerns. Scaling illuminates aspects of living forms, from the existence of lungs to (perhaps) the rate of our metabolism.

These four themes don't exist in isolation but can interact with and even depend on each other, as we'll see in the chapters to come. The precision of biological circuits often depends on the statistics of random motion. Random motion nudges the positions of biological components to facilitate their self-assembly. Self-assembly into larger structures is subject to the dictates of scaling laws. All these processes and principles together create the explanatory framework of biophysics.

· · ·

Understanding life brings with it the ability to influence life. This isn't in itself a new insight. Our knowledge of the immune system and the behavior of microorganisms, among other topics, has enabled us to triumph over a multitude of diseases that ravaged humanity in the past. In the twentieth century alone, for example, more than 300 million people died of smallpox, a disease that has now vanished thanks to the invention of vaccines. Our knowledge of genetics, biochemistry, and many other subjects lets us coax plants and animals to produce enough food for over seven billion people, four times as many as inhabited the planet just a hundred years ago. In recent years, we've learned how to alter organisms at their core, directly reading the information carried in genomes and rewriting it to modify form and function. As we'll see, these contemporary advances required taking seriously a biophysical view of life, acknowledging the tangible, physical character of DNA and other molecules to design tools that quite literally push, pull, cut, and connect life's pieces.

A biophysical perspective also helps us make sense of the implications of these new biotechnologies and the difficult choices they bring. We'll encounter, for example, methods to engineer the extinction of the mosquitoes that spread malaria, dengue fever, and other diseases, bringing to mind both the dismal legacy of human-induced extinctions and the uplifting histories of past eradications of disease. The decision whether to deploy such methods requires understanding how they work and how they differ from past tools. At a more personal level, our ability to read our own genetic code brings with it the prediction of likelihoods of various illnesses in ourselves or in our children; our nascent ability to edit genomes offers the chance to alter these likelihoods. What would it mean to alter the genome of an unborn child to try to avoid cystic fibrosis, or cancer, or depression? Whether to take such an action is both a deeply personal decision and one with serious ethical and societal implications. Making such decisions can, and should, be aided by an understanding of what genes, genomes, cells, and organisms actually are, and the processes that shape the relationships among them. As we'll see, the physical nature of life's materials, as well as fundamental issues related to randomness and uncertainty, influence what we can and cannot do with our new technologies.

· · ·

Our exploration of biophysical themes includes examples spanning the variety of life. We consider the normal workings of organisms, including ourselves, as well as the pitfalls of disease and the intersections of biology and technology. In part I ("The Ingredients of Life"), our journey begins inside cells. We delineate the pieces that make up living things, materials like DNA and proteins that also exemplify a sort of universality, as they make up every living thing ever discovered. The molecular characters in this first part of the story will likely be familiar from high school biology, but we focus on the physical traits that guide their functions. We find stiff strands of DNA, two-dimensional liquids that define cell boundaries, and three-dimensional sculptures made of single molecules. In part II ("Living Large"), we expand our horizons to look at communities of cells, including embryos, organs, and the consortia of bacteria that live inside each of us. We also explore scaling relationships that govern the shapes of animals and plants, revealing why an elephant can never be as athletic as an antelope. In part III ("Organisms by Design"), we return to the microscopic world of DNA, but now, having developed deeper connections between molecules and organisms, we tackle the genome. We learn what it means to read, write, and edit DNA, learn how nature itself pointed us toward the tools that make these feats possible, and examine the opportunities and challenges these technologies present for our future.

As interesting as these topics and examples may be, their cumulative effect is greater than the sum of their parts. Biophysics transforms the way we look at the world. At the end of *On the Origin of Species*, Darwin writes:

> There is grandeur in this view of life, with its several powers, having been originally breathed into a few forms or into one; and that, whilst this planet has gone cycling on according to the fixed law of gravity, from so simple a beginning endless forms most beautiful and most wonderful have been, and are being, evolved.

I hope to convince you that Nature has a grandeur even deeper than what Darwin discerned. Rather than a contrast between the fixed, clockwork laws of physics and the generation of endless and beautiful forms, the two are inextricably linked. We can identify the crucial "simple beginning" not as the origin of life, nor the formation of our planet, but as the primeval emergence of the physical laws that characterize our universe. The influence of these laws on life didn't end billions of years ago, but rather shaped and continues to shape all the wonderful forms around us and within us. To discern simplicity amid complexity and to draw connections between life's diverse phenomena and universal physical concepts gives us a deeper appreciation of ourselves, our fellow living creatures, and the natural world that we inhabit. I hope you'll agree.

PART I

The Ingredients of Life

DNA: A Code and a Cord

A beige gelatinous slab speckled with bacterial colonies hangs in the National Portrait Gallery in London. The bacteria contain copies of DNA from the artwork's subject, Nobel laureate John Sulston. Though Sulston's friends probably wouldn't recognize the likeness, the artist, Marc Quinn, notes that the piece "is the most realistic portrait in the Portrait Gallery" because "it carries the actual instructions that led to the creation of John."

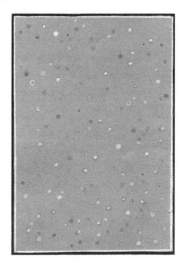

These days even small children are taught that DNA somehow makes you "you," setting the color of your eyes, the shape of your nose, your fondness for cilantro, and more. We've grown accustomed to the idea that DNA encodes instructions that govern us, but what does "encodes" actually mean?

We begin our biophysical exploration of the machineries and mechanisms of life with DNA; it is iconic and familiar, yet abstract in many descriptions we encounter. Understanding how instructions are embedded in DNA occupies us for a few chapters, as we introduce proteins, genes, and networks of interactions among these constituents. Scientists' conceptions of how these pieces form both a message and the means of reading this message have evolved over the past few decades, and DNA's intricacies are still being unraveled. Recent years have seen the development of breathtaking ways to manipulate the code contained in DNA, for ends not yet fully imagined, a subject we examine in part III. In this chapter, we focus on DNA alone, which already allows us to introduce connections between biology and physics, and between science and technology.

FOUR VIEWS OF DNA

More than just a set of abstract instructions, DNA is a substance with shape and structure whose physical properties are intimately tied to its functions. What *is* this substance? Is it solid or liquid, stiff or floppy, compressed or relaxed? DNA is multifaceted, and we can focus on different aspects of it depending on what we care about. Here are four views of DNA.

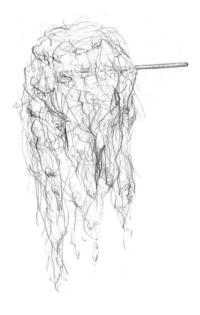

DNA is a colorless goo. We can hold DNA in our hands and see it with our unaided eyes. This isn't hard to do: a blender and some kitchen chemicals enable the extraction of DNA from a bowl of strawberries or a cup of peas. The recipe is roughly this: puree the fruits or vegetables in a blender, ripping their cells apart from each other. Add detergent to disintegrate cellular membranes. Sprinkle a dash of meat tenderizer or pineapple juice, supplying enzymes that digest proteins. DNA is now the only cellular component left intact. Add rubbing alcohol, which dissolves the protein pieces but not the DNA. The DNA clumps into long strands you can draw out with a toothpick, collecting a cloudy, stringy, white blob. That's DNA. It's not an awe-inspiring sight. I was once brought to tears extracting DNA during a classroom demonstration, but that was because I made the terrible decision to use onions as my source material—conveniently colorless, but a painful puree.

DNA is a code. At the other extreme of tangibility, we can think of DNA as an abstract code composed of four symbols. These symbols are often denoted by letters—A, T, C, G—but four colored squares work just as well. The particular sequence of symbols conveys information about how your cells should build what they need to build and do what they need to do. How much information? Let's compare it to the amount of digital information stored in a portable music player. These days we're used to thinking about "bits" and "bytes"—units of information. A bit (**bi**nary dig**it**) is anything that can have just two values: yes or no, 0 or 1, a magnet pointing north or south. A one-gigabyte thumb drive has about eight billion bits of information, *giga* meaning "billion" and a byte being eight bits. Its contents can be expressed as a particular string of eight billion ones and zeros (. . . 01110100110010100011 01110100001101110001101011 . . .). How many bits is each person's DNA sequence? Three billion symbols—that is, three billion As, Ts, Cs, and Gs—make up your DNA. We could make a dictionary like this, for example, to translate each symbol into a binary code:

00 = A
01 = T
10 = C
11 = G

A sequence like ATTGC would be equivalent to 0001011110. Our three-billion-letter genome, therefore, carries six billion bits of information—less than a gigabyte, and probably a small fraction of the storage capacity of the phone in your pocket. This presents a puzzle: I seem much more complex than my phone, despite my apparent paucity of information! We grapple with the concept of complexity through much of this book. For now, there's a more immediate question: How does this abstract picture of codes and information relate to the stringy blob illustrated previously?

DNA is a molecule. Like all molecules, DNA consists of atoms, in its case atoms of carbon, hydrogen, oxygen, nitrogen, and phosphorus, held together by chemical bonds. The four symbols of the code mentioned above are really four assemblies of atoms, called *nucleotides,* stitched together to form a long chain. The upper illustration depicts all the atoms in an adenine nucleotide (A), with carbon atoms as black, nitrogen as blue, oxygen as red, phosphorus as purple, and black lines denoting chemical bonds. (For clarity, I omitted the many hydrogen atoms.) The lower illustration shows the atoms in the four-nucleotide sequence ACTG, with the ellipses (. . .) indicating where this segment would connect to adjacent nucleotides if it were part of a longer strand. Specifying the sequence of nucleotides in the chain suffices to identify a DNA molecule—the shorthand of ACTG is exactly equivalent to the array of atoms I've drawn, and it can't refer to any other set of atoms than this.

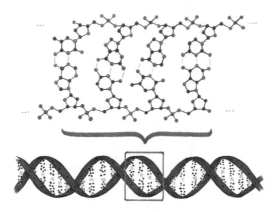

DNA is a double helix. The interactions between atoms determine the structure of a molecule, and the structure of the molecule governs its function. Any of the four nucleotides, A, T, C, G, can be linked to any other to form a single strand of DNA. But the nucleotides also interact *between* strands, though more weakly, in a specific way: A and T bind together, as do C and G. (We say that A and T are complementary, as are C and G.) A single strand of DNA, for example, AGCCTATGA, binds its complementary strand, TCGGATACT. The illustration shows the atoms in the double-stranded DNA formed from ACTG and its complement, TGAC, with the green lines indicating the interstrand bonds. Interactions among the atoms drive the two DNA strands to wind around each other like twisting ivy, forming a double helix. The illustration echoes the cartoon double helix we've all seen countless times; the smooth ribbons and well-ordered dots are a schematic of the more complicated arrangements of atoms and bonds in three-dimensional space.

The iconic double-helical form of DNA is functional as well as elegant. The two complementary strands convey redundant information: if I tell you the sequence of one strand, you know the other, since each nucleotide is the complement of its partner. This redundancy reveals how information can be transmitted from one cell to its two daughter cells as it divides: the DNA is unzipped and the complement of each original strand is synthesized, giving two DNA double helices from the original one.

James Watson and Francis Crick figured out the structure of double-stranded DNA in 1953, based on exquisite X-ray measurements performed by Rosalind Franklin and graduate student Raymond Gosling. (This story is a fascinating one, filled with cleverness and insight as well as ethical lapses and tragedy, and it is well told elsewhere; see the references.) Before this, no one knew what DNA molecules might look like. The most prominent hypothesis was from Linus Pauling, the chemist who first formulated the modern notion of a chemical bond, who suspected that it formed a three-stranded twisted fiber (a triple helix). The revelation of the double helix made it clear how DNA's structure might enable the transfer of genetic information by the copying of complementary strands. Other consequences of DNA's structure are less obvious, though, and we're still unraveling the mysteries of DNA today.

In living cells as well as in sterile test tube solutions, single strands of DNA will spontaneously wrap themselves into a double helix if their nucleotides complement each other. No external scaffolding or microscopic ropes and pulleys are required: the DNA contains within itself the mechanism of its organization, highlighting a theme of self-assembly that surfaces repeatedly in our explorations of life.

Each of the four views of DNA depicted above is useful, emphasizing attributes relevant to various roles. The fibrous goo extracted from pureed cells may be homely, but all of the more glamorous uses of DNA—plucking it from cancer cells to map the genes they carry, harvesting it from crime scenes to track down suspects, and more—must acknowledge the material, corporeal character of DNA if they are to work. As an abstract code, the sequence of symbols specifies the information carried by the molecule. When we say that your DNA is unique, it means that your sequence of nucleotides or colored squares is different from mine. (It's only slightly different; over 99% of our squares would match.) When we say we know the genome of some organism, we mean that we know the full sequence of symbols. This tells us a lot, but it also leaves a great deal undetermined, as we'll see. We might care about the atomic level of detail—the exact architecture of atoms making up DNA rather than just the symbolic code of constituents—if, for example, we're designing tools that cut DNA strands or splice them together, which we encounter in the context of genome editing in part III. More often, however, the arrangement of the units, the As, Cs, Gs, and Ts, suffices. The double-stranded helix describes how DNA is situated in space; the size, shape, stiffness, and electrical charge of double-stranded DNA govern its packaging in cells and the readout of information it contains. As our first illustration of the importance of the physical character of double-stranded DNA, let's look at a process that has transformed biotechnology: the polymerase chain reaction.

DOES DNA MELT?

Imagine you'd like to make an exact copy of some DNA. You could begin by separating the two strands of the double helix and then creating a new complementary strand for each of the resulting halves. That is, in fact, what your cells do every time they divide, with a particular protein machinery performing the initial unzipping of the double-stranded DNA. Outside of cells, we've developed another approach that allows us to replicate DNA at will, taking tiny amounts of DNA and transforming them into innumerable identical copies to yield enough material to run tests on or to transport to new targets. The tiny amounts of nucleotide strands could, for example, be swabbed from a crime scene to assess similarity to DNA from suspects; sampled from the amniotic fluid surrounding a fetus to search for genetic abnormalities or the presence of bacterial or viral DNA; excised from a tumor to map mutations in the nucleotide code indicative of cancer; or extracted from a Nobel Prize winner to be replicated and reinserted, in fragments, into bacteria that make up an artwork. Just as in natural replication, the artificial replication of DNA relies on drawing the double helix apart, which in turn relies on the physical phenomenon of phase transitions.

Rather than asking how we might separate the two strands of a DNA double helix, imagine asking how we could separate the tightly connected water molecules that make up an ice cube. We know the answer: we'd heat it. Above 0°C (32°F) ice melts into liquid water, in which each water molecule meanders around, only transiently bound to any other molecule. In general, temperature is the nemesis of attraction and order, a recurring theme in physics. For water, the transition between solid and liquid is sharp, occurring precisely at the "melting temperature," which is 0°C at standard atmospheric pressure. Even a few degrees below 0°C, water is solid; even slightly above, it's liquid. This transition isn't sharp for all substances, however. If you heat honey, it gets progressively less viscous (flowing more easily) as the temperature rises, rather than suddenly changing at one particular temperature.

Recalling that the bonds between the two strands of a DNA double helix are weaker than the bonds within a strand, we might expect that we could use heat to separate DNA strands without destroying them, and this is, in fact, the case. But is the transformation sharp or smooth? In other words, does DNA have a melting transition? The answer is important if we want separation to enable replication. If there *is* a melting transition, we'd be assured that by raising the temperature, even to just a few degrees above the transition, we'd have complete separation (upper illustration). If there isn't a melting transition, we'd likely have some unseparated DNA that couldn't be copied (lower illustration).

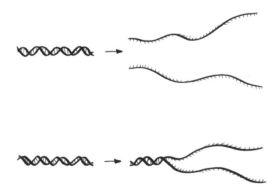

In the latter case, though we could keep heating until every last segment is completely separated, this would in practice likely require such high temperatures that the DNA itself, and any other biological molecules that are present, would be damaged.

It turns out that, for double-stranded DNA, the transition into single strands is a sharp one. DNA *does* melt. If we take a beaker containing DNA and heat it, the molecules are double-stranded up to a specific melting temperature, and just above this they separate into single strands. Not only can we measure this in the lab, but we understand its origins. Making sense of phase transitions—transformations between liquid, solid, and gas, or between magnetic and nonmagnetic forms, or between any of the different structures that materials might take—was one of the great triumphs of twentieth-century physics. The melting of DNA is a transition from order to disorder that maps onto other transitions, though with subtleties of its own.

All phase transitions reflect a conflict between order and disorder. Order is typically driven by the energy associated with attraction or alignment, while disorder is often driven by geometry, the ways in which constituents can arrange themselves in space. Temperature amplifies the impact of disorder. At low temperatures, the tendency to be ordered wins; at high temperatures, disorder is dominant. Water molecules, for example, are marshaled into the crystalline arrangement of ice when cold; if warmer, they succumb to the randomness of being liquid. Our statement that the melting transition is sharp means that there's a specific temperature separating these two outcomes and therefore a clear demarcation of the ordered and disordered phases.

The energy of ordering and the varieties of disorder both depend on the dimensions the substance can explore. The consequences for phase transitions are dramatic: in general, theory predicts that one-dimensional materials shouldn't exhibit sharp transformations between phases. Water molecules arranged in a row, for example, wouldn't melt at a specific point; disorder would emerge at even the coldest accessible temperature and would grow steadily as the temperature increases.

To a good approximation, the long chain of double-stranded DNA is one-dimensional—rungs on a ladder, one after the other. Its sharp

melting, readily observed in the lab, therefore seems to confound expectations. As strands come unbound, however, the liberated, flexible,
single-stranded DNA bends and twists through three dimensions, subject to random forces common to all molecules whose character we explore further throughout this book. Though the motion is random, its
consequences are robust and predictable—another recurring theme—
as the resulting configurational freedom grants the overall separation
a precise transition temperature, which is generally the case for
three-dimensional materials. With experimental data and theoretical
understanding in hand, we can even predict, for a given DNA sequence, the temperature at which it will come apart. These transition
temperatures are typically around 95°C (200°F), a bit below the boiling
point of water, with the exact value depending on the particular nucleotide sequence.

We can therefore separate DNA in a test tube just by heating it. The
next task for our goal of replicating DNA is to create the complements
of each strand. We could borrow the machinery that your cells use,
called *DNA polymerase*, but these proteins would form a dysfunctional,
rubbery glob, rather like a boiled egg white, at the temperatures necessary to melt the DNA. (Egg whites are, in fact, mostly protein.) There
is, however, a clever way around this: We use DNA polymerase from
bacteria that live in hot springs, creatures whose proteins have evolved
to run smoothly when hot. Like all DNA polymerases, these require a
little bit of double-stranded DNA to get started with their replication
of a single strand. These ingredients were all in hand by the early 1980s;
bacteriologists discovered and purified the hot-springs polymerase,
for example, in 1976. In 1983, the recipe that combines nucleotides,
polymerases, temperature, and DNA came to scientist Kary Mullis

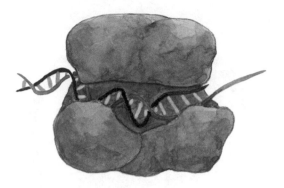

during a late-night drive through California's Coast Range, along with the realization that this would enable simple and almost limitless replication of DNA. The process is now known as the *polymerase chain reaction*, or PCR.

To perform PCR, we first gather the ingredients, dissolved in a salt-water solution: the little bit of DNA we wish to replicate, DNA polymerase proteins, an abundance of individual nucleotides (As, Cs, Gs, and Ts), and a lot of *primers*—segments of DNA just a few nucleotides long that are complements of the ends of the DNA to be copied.

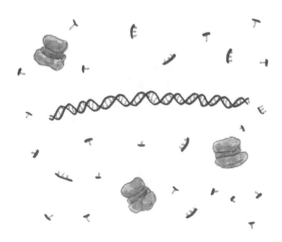

Next, we raise the temperature to around 95°C (200°F) so that the DNA melts, giving two single strands from the original double helix.

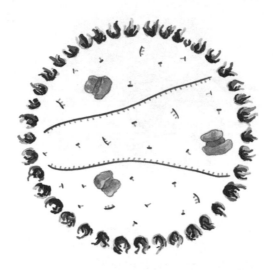

We next lower the temperature, so that the primers will bind to the ends of the single-stranded DNA. There are a lot of primers, so it's much more likely that the single-stranded DNA encounters a primer than its former partner strand; we put predictable randomness to work for us. The polymerase will then bind and craft the complementary DNA strand. When it finishes, we have two double-stranded DNA helices, from our original one.

We then repeat. With one more cycle of warming and cooling, we end up with 4 DNA fragments; at the next round 8; then 16, 32, 64, ...

Ten doublings give 1024 pieces of DNA; twenty give over a million; thirty repetitions (not difficult with an automated heating and cooling machine) give over a billion!

We can therefore take a little bit of DNA and turn it into a *lot* of identical DNA, amplifying our original whisper into a chorus of replicas to be used in whatever diagnostic, forensic, or therapeutic application we like. In living things, copying DNA requires the elaborate dance of reproduction, creating an entire new cell or even a new organism. With the polymerase chain reaction, we can copy DNA at will. The recipe is beautifully simple. As Mullis himself wrote, the "almost universal first response" of molecular biologists upon learning about his invention was, "Why didn't I think of that?" Mullis continues, "And nobody really knows why; surely I don't. I just ran into it one night."

By replicating DNA, we can create enough copies to "sequence" it— in other words, to determine its particular pattern of nucleotides—using techniques that we describe in part III. With this approach, we've mapped the genomes of humans and many other organisms.

You may wonder, since PCR needs primers that bind to the DNA of interest, creating a short double-stranded span, don't we have to know the sequence we're amplifying before we even begin? Not quite. First of all, for many applications we do know some part of the genome of the organism whose DNA we're working with, and can design primers that bind to this part. More generally, though, we can cut the unknown DNA into fragments and link them to DNA that we know well, for example, the genomes of easily grown bacteria, using naturally occurring proteins that sew DNA strands together. Then primers to the known part of the DNA can seed the progression of DNA polymerase onto the unknown parts. It's with this approach, for example, that the genome of the woolly mammoth, an animal that's been extinct for several thousand years, was sequenced from ancient remains.

The polymerase chain reaction is essential to nearly all of what we do with DNA, and it is made possible by bringing together quirks of biology (like heat-loving microorganisms) and generalities of physics (melting phase transitions), each of which on their own may have seemed tangential to practical aims. PCR also highlights a message central to many modern technologies as well as natural processes: DNA is more than a code or an abstraction, but is a tangible, physical object. Especially with notions of self-assembly and predictable randomness in hand, we can understand and even invent methods to work with this crucial molecule.

There's a lot more we can, and will, ask about DNA: How much do you have? How is it stored, organized, and deciphered by your cells? How could you alter the code written into your (or your unborn child's) genome? These questions tie together physical properties and biological functions. To answer them, though, we need to introduce another key player on the cellular stage: proteins. In the next chapter, we see what a protein is and examine the interplay between proteins and DNA.

2

Proteins: Molecular Origami

At the core of nearly every action, every task, and every event in your body is a protein. Proteins in red blood cells soak up oxygen from the air you breathe. Proteins tug on other proteins to contract your muscles. Proteins extend and retract protrusions with which immune cells squeeze through your tissues. Proteins in your eyes capture light and trigger electrical impulses, while other proteins open and close gates that send these impulses to your brain. Many kinds of proteins are inside every cell, and many are outside as well, making up, for example, the elastic matrix of your flesh. So: What *are* proteins?

Like DNA, a protein is a molecule composed of a chain of simple units. In DNA, these units are any of four nucleotides. In proteins, these units are any of 20 amino acids. Double-stranded DNA, regardless of its nucleotide sequence, adopts a double-helical structure. In contrast, proteins have structures that are determined by their particular amino acid sequences. Every distinct protein has a different pattern of amino acids and therefore a different three-dimensional shape. The blueprints for the structure and the tools for its construction are encoded within the protein itself. Proteins provide perhaps the most striking manifestation of the theme of self-assembly, nature's encoding of instructions for organizing matter within the matter itself, to be activated and realized by universal physical forces. Though self-assembly isn't unique to living things—sand piles, for example, arrange themselves into cones tilted at specific angles and soap bubbles form themselves into spheres—it is ubiquitous in biology. Considering proteins, we'll see how forces generate forms, how the process usually succeeds but sometimes catastrophically fails, and how computers struggle with geometric calculations that molecules perform in microseconds.

PROTEINS IN THREE DIMENSIONS

An amino acid chain in water bends, twists, and folds into a particular shape. Two very common motifs found in proteins are helices and sheets.

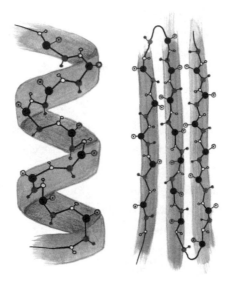

I haven't drawn all the atoms in these structures—just a few representative dots and the bonds between them. Helices and sheets are so common in protein structures that we often depict only stylized shapes—a smooth helix about one nanometer in diameter (one billionth of a meter), and a sheet or ribbon about a third of a nanometer wide—rather than showing every component atom.

The first protein to have its three-dimensional structure revealed was myoglobin, a protein that carries oxygen in muscle, in 1958. As with DNA and many other molecules, this was made possible by illumination with X-rays together with mathematical analysis of the resulting intensity patterns, in this case performed by a team led by John Kendrew at the University of Cambridge. X-ray imaging requires proteins to solidify into crystals, similar to the sugar crystals you might make in your kitchen, and even now there is an art to coaxing proteins to crystallize. Kendrew's team struggled with myoglobin from porpoises,

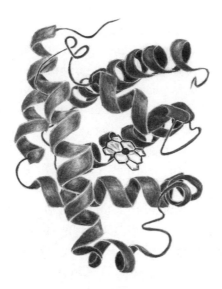

penguins, seals, and other creatures before stumbling on sperm whale meat, conveniently on hand in the freezer of Cambridge's Low Temperature Research Station. (Deep-diving, air-breathing sea creatures have high concentrations of myoglobin in their muscles, allowing them to

store more oxygen and surface less often, hence the focus on these animals.) The sperm whale protein formed "the most marvelous . . . gigantic crystals." From these, Kendrew and his team were able to determine that the 153-amino-acid chain of myoglobin folds into a structure composed of eight helices and some nonhelical spans that together hold onto a flat compound containing a single iron atom that binds oxygen (upper illustration, page 31).

We can again turn to the marine world for an example of a protein composed mainly of sheets. Green fluorescent protein, commonly called GFP, is a light-emitting protein first found in bioluminescent jellyfish. GFP is a chain of 238 amino acids, folded into a barrel of sheets about 3 nanometers wide that surrounds the part of the molecule responsible for color (lower illustration, page 31). The protein has become much more than an oceanic curiosity. GFP has been engineered into bacteria, fungi, plants, and even animals as diverse as fruit flies and zebrafish, in which it serves as a beacon allowing researchers to visualize particular types of cells and see how they grow, move, and divide. GFP can also be fused to other proteins, creating reporters that reveal where in cells these proteins are, how they behave as the cells perform various tasks, and how they associate with other proteins to form more complex architectures. A rich palette of fluorescent proteins now exists, emitting light in a rainbow of colors, derived from GFP or other proteins discovered in corals, with names ranging from the dull ("red fluorescent protein") to the more evocative ("tangerine," "cherry," "plum"—a whole series of fruit names). The ensemble has made possible multicolor imaging of the machinery of living things, with applications far removed from these proteins' marine origins.

PROTEIN PORTRAITS

The three-dimensional structure of a protein matters: it's intimately tied to the protein's chemical or physical tasks. In GFP, for example, the barrel shields the light-generating unit from water and dissolved oxygen that would quench its glow. A few more examples make the relationship between structure and function even more evident.

Thin membranes separate spaces within a cell and also demarcate the cell's inside and outside. Specific proteins embedded in membranes, often joined together into barrels or rings, transport atoms and molecules across. Ion channels are one class of these transporters, allowing specific charged atoms (ions), such as potassium, sodium, and chlorine, in and out of cells through a central pore that can be open or closed. Controlling flows of ions is a crucial task. The motion of your eyes

scanning this page and the thoughts racing through your brain are both manifestations of the electrical voltages generated by the re-distribution of ions across membranes. Many toxins produced by animals such as snakes and scorpions work by interfering with ion channel proteins, shutting down the nervous system of their victims. The illustrated example on page 33 is a potassium channel, drawn end-on with the membrane, not shown, in the plane of the page. The red dot is a potassium ion, traveling toward or away from us and therefore entering or exiting a cell. The channel is actually composed of four identical protein molecules that loosely bind together to construct a membrane-spanning pore.

Channels can open and close, but other proteins can perform more elaborate gymnastics. I've depicted an assembly of two molecules of a protein called kinesin (lower illustration, page 33). As the name implies (think "kinetic"), it's involved in motion. Each kinesin protein takes the shape of a long stalk with a bulbous end, connected by a flexible amino acid joint. The two stalks bind together and can adhere to cargo that needs to be ferried from one place to another, for example, packets of chemicals synthesized deep in the interior of a neuron and stored for release near its edge. The whole complex can then walk along tracks within a cell. "Walking" isn't metaphorical: the two feet alternately bind and unbind from the tracks, ambling foot-over-foot to reach their destination. (Traditionally, these feet are called *heads* and the foot-over-foot motion is called *hand-over-hand* motion. Yes, the nomenclature is puzzling.) The tracks themselves are also made of proteins, arranged into rigid filaments; again, their three-dimensional shape allows them to play their role.

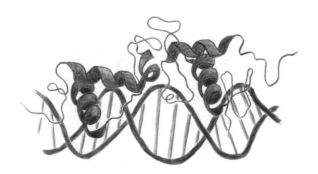

The structure of proteins influences their interactions with each other and with other molecules they encounter, such as DNA. Many proteins bind to DNA to guide the readout of its genetic information, as we explore in more detail in the next two chapters. These DNA-binding proteins must adopt a shape that conforms to the curves of the DNA double helix. Amino acid helices can nestle in the DNA grooves and are a common motif in these proteins' structures, illustrated here for a hormone-sensing molecule known as the glucocorticoid receptor. (A pair of these proteins works together; I've drawn only the DNA-adjacent regions of such a pair.) When the receptor encounters and latches onto a hormone called cortisol, its structure changes, and only then is it capable of binding to DNA, setting in motion a series of events that, among other things, inhibit the organism's inflammatory immune response. You've probably encountered cortisol in ointments, where it's often called hydrocortisone, and made use of its activation of receptor proteins to reduce the redness, itch, and swelling of your body's reaction to poison ivy, insect bites, and other irritants.

PROTEIN FOLDING

As we've seen, protein structure is intimately connected to protein function. A protein doesn't begin its life fully formed, however. Every protein is made by cellular machines that attach one amino acid to the next, sequentially, like paper clips linked together into a chain. There isn't any scaffold that gives structure to the string-like molecule, arranging it into stacked sheets, tangles of helices, or any of a near-infinite variety of possible forms. Rather, the protein *sculpts itself* into its proper shape. The amino acid sequence of the protein encodes the determinants of its structure; the protein self-assembles.

Each of the 20 amino acids has particular physical properties. Some have positive electrical charge; some are negative; some are neutral. Some are large; some are small. Some are greasy ("hydrophobic") and prefer to separate from water; some are "hydrophilic" and mix well with water. Imagine a protein with several positive amino acids in a row (red circles in the illustration), followed by a string of neutral hydrophilic amino acids (black circles), and then several negative amino acids (blue circles). Opposite charges attract; so left to its own devices, the protein folds to bring together the contrasting ends.

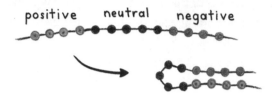

Or imagine a protein with amino acids that are hydrophobic (orange squares) and hydrophilic (black circles). The protein is surrounded by water—water makes up the majority of the cellular interior—and will fold to bury the hydrophobic bits in the center to be surrounded by their water-loving colleagues.

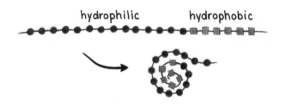

I've drawn these schematic illustrations in two dimensions for clarity. Really, you should imagine a roughly spherical core of hydrophobic amino acids surrounded by a shell of hydrophilic amino acids.

In any real protein, many such interactions between amino acids and each other and between amino acids and the surrounding water take place, giving rise to forces that pull the protein toward a particular conformation. Every protein is synthesized in the cell as a chain of amino acids, and that chain folds itself into its optimal three-dimensional shape. The technical term is, in fact, *protein folding*.

As with nearly everything in biology, this blunt picture isn't quite true. Some proteins, especially large ones that tend to aggregate, need a bit of help to fold, and a group of other proteins called *chaperones* comes to their aid. Assemblies of chaperone proteins contain chambers that protect the nascent protein from the complexities of the crowded cellular environment, facilitating the proper folding of the amino acid chain. Chaperones notwithstanding, the general idea of proteins possessing the plans for their own architecture is powerful and prevalent throughout the living world.

Every protein we've described above, and tens of thousands of others, folds within a fraction of a second into a three-dimensional form, bypassing the innumerable pitfalls and dead ends of shapes that don't quite satisfy the interactions their component parts prefer. This is an amazing feat—like a piece of paper spontaneously folding itself into a perfect origami sculpture. What's more, for the vast majority of proteins the sculpture is uniquely determined by the amino acid sequence. In other words, a given sequence always folds into the same shape. Every green fluorescent protein forms a barrel; every myoglobin folds into the same collection of curls.

A few illustrations might help convey how remarkable this self-organization is. Imagine a sequence of amino acids that, as usual, has positively charged, negatively charged, neutral hydrophilic, and hydrophobic amino acids. (Charged amino acids are always hydrophilic, by the way.) The chain could fold into the arrangement on the left, which is pretty good—the hydrophobic pieces are buried in the interior and opposite charges lie next to each other. But the arrangement on the right, of exactly the same sequence, is also pretty good, for the same reasons.

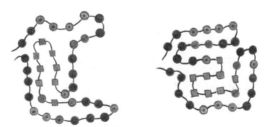

The two conformations are certainly different. We might imagine that if this protein needs to bind to some small molecule—a hormone, for example—the presence of a "pocket" in the first form but not the second would make the first functional and the second useless.

Understanding how an amino acid chain adopts a single, optimal shape turns out to be remarkably puzzling. For a random amino acid

sequence—imagine blindly picking amino acids out of a hat and stringing them together—analysis of forces and energies reveals that we should expect a huge number of "pretty good" configurations, far too many to expect the chain to routinely find a unique endpoint to its folding. Nature sidesteps this multiplicity of possible forms; the proteins that actually exist in the real world aren't random, but rather are those that have been selected by four billion years of evolution. Organisms that encode amino acid sequences that don't fold into unique shapes are plagued by dysfunctional and perhaps even harmful proteins, and hence are less likely to live and reproduce. Those that persist are those that encode amino acid sequences with a clear, unique path to three-dimensional structure.

The result of all this, as we've seen, is the general principle of a one-to-one correspondence between amino acid sequence and protein structure for the proteins actually present in ourselves and other organisms. If we know the structure of one kinesin molecule, we know the structure of every kinesin molecule. If we know the structure of one cortisol receptor, we know the structure of every cortisol receptor. As with all rules of thumb, however, there are exceptions, and the exceptions to this one are exceptionally important.

One set of rule-breakers are the "intrinsically disordered proteins," which don't have a particular form at all. Examples include some of the proteins that make up the pores of the membrane surrounding cell nuclei. It's thought that the spaghetti of disordered proteins occupying the pore provides flexibility for the transport of different-sized objects into and out of the nucleus.

More interesting, in my opinion, are the proteins that have *a few* stable configurations—not one unique form and not an amorphous blur, but rather a couple of architectures that they might toggle between, like a light switch toggles between stable on and off positions. In the past few decades, we've discovered not only that proteins like this exist but that they contribute to some perplexing diseases. They also provide a warning, in case you needed it, not to indulge in cannibalism.

KURU AND CANNIBALISM

In Papua New Guinea in the 1950s, an epidemic of a strange disease characterized by tremors and uncontrollable fits of laughter struck villages of the Fore people, killing up to 200 people a year among a total tribal population of about 11,000. (For scale, imagine 150,000 New Yorkers dying eerie deaths every year.) From patterns of illness and contagion, anthropologists and medical researchers deduced that the disease, named *kuru* after the Fore "to shake," was spread by ritual cannibalism among the Fore: when a person died, family members consumed the body, an act of love and respect that helped liberate the spirit of the deceased. The Australian government that was ruling Papua New Guinea banned cannibalism around this time, which led to a steady decline in the prevalence of kuru. It took decades, however, to figure out the actual cause of the disease. The culprit was not a bacterium, a virus, or a parasite but rather a protein—an unusual protein that doesn't have one unique structure but can adopt one of two forms. In the "normal" form, the protein performs its usual functions. In the "misfolded" form, it does not. Even more perniciously, however, the misfolded protein induces others to adopt its aberrant shape and join with it to form fibrous aggregates. In this way, the deviant protein is infectious: ingesting proteins of the misfolded form, some of which make their way to the brain, induces structural changes in molecules with an otherwise benign amino acid sequence. This change amplifies itself through the victim's nervous system and is propagated further still if the victim dies and is eaten by another villager. The chain of events is reminiscent of Kurt Vonnegut's novel *Cat's Cradle*, in which the fictional "ice-nine" form of water is solid at room temperature and, upon contact with the normal form of water, induces its crystallization and transformation into more ice-nine. The resulting chain reaction is even more deadly than kuru. Unlike ice-nine, however, kuru is real.

Proteins that can fold into multiple forms and that act as infectious agents are called *prions*, and we now realize that they drive several diseases of humans and other animals. One of these is bovine spongiform encephalopathy, more evocatively known as mad cow disease. Like kuru, it is neurodegenerative, inducing tremors, excitability, and poor motor coordination, but in cattle rather than humans. An outbreak in the United Kingdom in the late 1980s infected about 200,000 cows, and over 4 million animals were killed to halt the spread of the epidemic. Before it was stopped, however, it spread to humans. Well over a hundred people died from the human analog, called variant Creutzfeldt-Jakob disease, almost certainly from eating diseased animals. How did the animals get infected? Cannibalism! Meat-and-bone meal, believed to enhance the animals' growth and productivity as well as providing a use for leftover parts, was commonly fed to farm animals. This livestock cannibalism was banned following these outbreaks, in the United Kingdom in 1989 and by now in most of the world, at least for ruminants like cows and sheep. (Meat-and-bone meal is, in general, still allowed as feed for other farm animals, such as chickens and pigs.)

The very existence of prions was contentious for quite a while. In the 1980s, researchers led by Stanley Prusiner at the University of California at San Francisco, following a decade of work, isolated the infectious agent of scrapie, the sheep analog of bovine spongiform encephalopathy, and identified it as a protein. Their announcement was met with intense skepticism—there's a sense of agency to the actions of bacteria, viruses, and parasites that a simple amino acid chain lacks, and it's understandably hard to imagine it propagating, amplifying itself, and causing disease. Nonetheless, painstaking analyses and elimination of other possibilities established the reality of the prion hypothesis.

Kuru and mad cows aside, prion or prion-like proteins pop up in other major diseases. Most notably, Alzheimer's disease is often accompanied by aggregates of misfolded proteins that resemble those of prion diseases. These aggregates don't seem to be infectious, though; transferring them from diseased to healthy animals doesn't transfer the neurological symptoms. What the causes and consequences of these protein agglomerates are remains unclear. More generally, there's still a lot left to learn about protein folding and misfolding.

PREDICTING PROTEINS

Returning to the overwhelming majority of proteins that *do* have a unique three-dimensional form, it remains surprisingly difficult to predict what that form will be given the protein's amino acid sequence. Such predictions would be immensely useful. Assessing how a potentially therapeutic drug would bind to a range of different proteins, for example, would be easier with the three-dimensional structure of each of these molecules in hand. Though structure determination has progressed considerably since our first glimpses of sperm whale myoglobin, it remains difficult, time-consuming, and capricious. The workhorse method of probing proteins with X-rays requires first coaxing them to form crystals, a craft that demands a great deal of trial-and-error tinkering, followed by interrogation with high-powered X-ray sources. Other methods exist, involving electron microscopes, for example, but none are fast or simple. It is appealing to think that, instead of actually making and measuring a protein, we might simply calculate, given the amino acid sequence, the three-dimensional structure it would adopt. Determining the amino acid sequence is easy thanks to the nature of the genetic code embedded in our DNA, which we elaborate in the next chapter. In principle, since we understand the physics of electrical forces and hydrophobic and hydrophilic interactions, we should be able to simply plug the amino acid sequence into a computer program, easy enough to write, that grinds through the requisite calculations, stopping when it has found the optimal folding of the molecular chain. In practice, however, the number of possible configurations is so enormous that even the fastest computers struggle to explore them all.

There have been lots of clever approaches devised to tackle this computational challenge—some focused on improving the algorithms for calculating forces and energies, some developing simplifications such as grouping sets of atoms together, and some exploring unconventional computer architectures. One could, for example, design a computer whose integrated circuits are tailor-made for calculating the sorts of forces experienced by amino acids, rather than the general-purpose integrated circuits of typical computers. This has been the approach taken by David Shaw, who focused the fortune he made as an investment manager to commission bespoke supercomputers devoted to the biophysical challenge of protein folding. Or one could use normal computers, but arranged in a large and haphazard array. This has been the approach of the authors of the "folding@home" program that runs in the background of volunteers' computers (anyone can sign up), using their otherwise idle moments to distribute calculations across tens of thousands of machines. Or one could try to recruit human minds. This has been the approach of researchers at the University of Washington who created a free protein folding game, called "foldit," in which players move amino acids like puzzle pieces on a screen, with the game's outcomes conveyed to the researchers. Or one could use artificial intelligence, training a computational neural network to infer patterns from known protein structures and apply them to predict new forms. This has been the approach of DeepMind, a company affiliated with Google, whose stunning performance placed it at the top of the 2020 "Critical Assessment of Protein Structure Prediction" contest. All these strategies and more have proved useful, though a quick and general method for computing the structure an amino acid chain will adopt remains elusive.

It is humbling to consider that the proteins themselves have solved the protein folding problem, shaping their structures within fractions of a second in every cell of every creature on earth. Self-assembly is awe-inspiring; it allows form to emerge from the pieces and forces intrinsic to nature's substances themselves. We'll see why this emergence can be so rapid and robust in chapter 6 when we discuss molecular randomness. But first let's explore the connection between proteins and DNA, defining the notion of the gene and setting the framework for revealing how self-assembled structures can form decision-making circuits in cells.

Genes and the Mechanics of DNA

We've called DNA a code, but what is it a code *for*? We've glimpsed a few of the many proteins an organism can create, but what determines this repertoire? The same concept, that of the gene, provides answers to both questions, binding together the abstract idea of biological information with the physicality of biological molecules. The power of genes as well as their limitations are inextricably linked to the physical properties of DNA, proteins, and the environment in which they exist. Discussions of genetic diseases don't often invoke the challenges of bending DNA or stuffing molecules into small spaces, but, as we'll see, such visceral concerns are important for making sense of how life works. Self-assembly is again central in this chapter as DNA and proteins, for example, join together to package genomes. Our other themes of predictable randomness, scaling, and regulatory circuits are also reflected in the handling of our genetic material, as our cells tackle challenges of size, shape, and disorder to organize their DNA.

WHAT IS A GENE?

We've seen that a protein is a sequence of amino acids stitched together with chemical bonds. The nucleotide sequence of the cell's DNA specifies each amino acid sequence. Groups of three nucleotides encode one particular amino acid. The DNA sequence TGG, for example, encodes the creation of one tryptophan, a hydrophobic amino acid. Both CGT and CGC specify the positively charged amino acid arginine. The sequence TGGCGT therefore indicates tryptophan linked to arginine. There isn't, however, a machinery that directly reads the DNA code and makes the corresponding protein. A molecule called *RNA* (ribonucleic acid) acts as an intermediary.

RNA, as its name might suggest, is similar to DNA. RNA is also a chain composed of any of four nucleotide units, three of which (A, C, and G) are the same as those in DNA and the fourth of which (U, uracil) is similar to DNA's T (thymine). A protein machine called *RNA polymerase* binds to a "promoter" sequence of DNA and then steps along the double helix like the slider on a zipper, spreading the two strands apart, reading the nucleotide sequence of one of the strands, called the *template strand*, and constructing a single-stranded chain of RNA. The process, which copies information from one form (DNA) to another (RNA), is called *transcription*, analogous to the transcription of spoken words into text or handwritten notes into type.

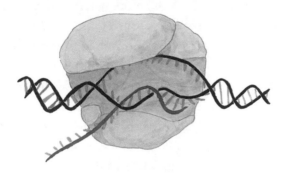

The RNA is complementary to the DNA template strand and is therefore identical in sequence to the other strand of the DNA double helix, called the *coding strand*, except for the Ts being Us. So, for example, the DNA coding sequence ATCGTT, mirrored as TAGCAA on the template strand, would be transcribed as the RNA sequence AUCGUU. Another machine, called the *ribosome*, translates the RNA strand into protein. The ribosome travels along the RNA, interacting with each three-nucleotide segment and attaching the appropriate amino acid to the protein it's constructing. The RNA sequence UGG, for example, encodes tryptophan; CGU and CGC both encode arginine. Some sequences (UAG, UGA, UAA) encode a "stop" message, telling the ribosome to halt its protein synthesis and detach from the RNA. The sequence AUG means "start here."

A particular segment of DNA therefore determines a protein that will be created via the processes of transcription (into RNA) and translation (into protein). Because DNA is transmitted from parents to children via sperm and egg, each such DNA segment enables the hereditary transfer of traits—the activities or characteristics of these particular proteins. Your ability to see color, for example, is made possible by three different proteins that react to different wavelengths of light, each produced in one of three types of cone cells in your retina. Differences of even a single three-nucleotide group, encoding 1 amino acid out of the approximately 350 in each of these proteins, can lead to subtle but measurable shifts in color perception. More dramatically, lacking the DNA sequence for an entire light-detecting protein results in one of several forms of color blindness.

One might think that these segments of DNA that encode proteins are what we mean by the term *gene*. This is almost, but not quite, correct.

Cells must not only specify the identity of the proteins they create but also control when this creation occurs and how much protein is made. Some stretches of DNA do not encode a protein sequence but rather affect whether other segments are read by the machineries of transcription and translation. For example, a class of proteins called *transcription factors* can bind to the promoter region at or near the starting point for RNA polymerase, diminishing or enhancing the likelihood that RNA polymerase will assemble and begin the transcription of DNA into RNA. We've seen an example of this already in chapter 2's glucocorticoid hormone receptor. Or a segment of DNA can be transcribed into RNA without subsequent translation into a protein, and this RNA can itself interact with DNA or other RNA molecules to influence protein synthesis. There are many ways in which RNA helps regulate the activity of cells, our understanding of which has in large part only arisen recently; RNA's status has been propelled from being merely a messenger between DNA and protein into a vital participant in these molecular conversations. An RNA called "Growth arrest-specific 5" made by cells sensing starvation, for example, attaches to the DNA-binding region of the glucocorticoid hormone receptor and thereby thwarts its recognition of its target DNA; the structural similarity of RNA and DNA allows the RNA to serve as a decoy.

Regulation of the processes by which genetic information is transformed into specific molecules is as important as the information itself, and it also enters the definition of what a gene is: A *gene* is a span of an organism's DNA sequence that encodes a particular, single hereditary characteristic, typically corresponding to a single protein or RNA sequence, and including noncoding regulatory sequences. It's a clunky definition, and one that's constantly changing, but life needn't comply with our desire for simple terminology. To further complicate matters, the term *gene* is still often used to mean "protein-coding DNA segment," its simpler and older meaning. Here I'll try to be transparent. The question we're now ready to ask is, thankfully, simple enough.

HOW MANY GENES DO YOU HAVE?

We can now read the genomes of all sorts of organisms, in other words, the complete sequence of As, Cs, Gs, and Ts. Because we can deduce the promoter sequences that instruct the transcription machinery to start and the terminating sequences that indicate stops, we can count the number of genes. For bacteria, we find a few thousand; each bacterium can generate roughly a few thousand distinct proteins. The bacteria that cause tuberculosis and cholera each have about 4000 genes in their genome, of which about 98% encode proteins. The genome of the *Lactobacillus delbrueckii* subspecies commonly employed to turn milk into yogurt has about 2000 protein-coding genes.

The human genome contains about 20,000 protein-coding genes. The number of noncoding genes, yielding RNA that doesn't go on to be translated into an amino acid chain, is harder to determine precisely but is estimated to be similar, around 20,000. Before you get too excited about your superiority over bacteria, because 20,000 is more than a few thousand, notice that your margin of victory isn't particularly large—there's less than a factor of 10 separating what most would consider an enormous difference in the complexity of the organisms. What's more, even among eukaryotes (organisms whose cells enclose their DNA in a membrane-bound nucleus), we're not very special. The common house mouse also has about 20,000 protein-coding genes, as do the Western clawed frog *Xenopus tropicalis* and the domestic horse. Some organisms have fewer genes. The fruit fly *Drosophila melanogaster* and the mushroom-forming fungus *Schizophyllum commune* each encompass about 13,000 protein-coding genes, and peregrine falcons about 16,000. The bread mold *Neurospora crassa* and the soil-dwelling amoeba *Dictyostelium discoideum* have about 10,000 and 13,000 protein-coding genes, respectively. Some organisms have many more genes than we do. The genome of the tiny water flea *Daphnia pulex*, about a millimeter long and nearly transparent, contains 31,000 protein-coding genes—the record, so far, among animals with sequenced genomes. Rice has about 30,000 protein-coding genes and maize (corn) about 40,000—*double* the human count—along with a few tens of thousands of noncoding genes. The number of genes tells us very little about the complexity or capabilities of organisms.

HOW BIG IS YOUR GENOME?

We've discussed your genome as a database of 20,000 protein-coding genes, but it's also a physical object, a series of A-T and C-G nucleotide base pairs that are the rungs of DNA's double-helical ladder, taking up physical space. Let's consider nucleotides first, then actual space. Your genome consists of about 3 billion base pairs. Most bacteria have much smaller genomes, typically a few million base pairs. The bacteria behind tuberculosis and cholera each have 4 million base pair genomes, and *Lactobacillus delbrueckii*'s genome size is about 2.3 million base pairs. But again, humans aren't especially remarkable or extreme in genome size. The mouse genome is similar in size to yours, while the fruit fly's is about 25 times smaller. The rice genome is also smaller, at about 430 million base pairs. (If this seems puzzling given the large gene count we stated earlier, don't worry—we'll return to this shortly.) Salamanders have especially sizable DNA, with genomes spanning 14 to 120 billion base pairs. The lungfish genome is 130 billion base pairs, and that of the flowering plant *Paris japonica* is a whopping 150 billion base pairs, 50 times larger than the human genome, making it the likely record holder for size. The single-celled amoeba *Polychaos dubiu* may surpass it with 670 billion base pairs, but this is somewhat controversial because its length was determined with outdated methods. (I'm amazed that no one has revisited this creature's DNA. If you're reading this and have a DNA sequencer and some spare time, go for it!) As with genes, there's no straightforward connection between genome size and the complexity of the organism.

Quantifying the numbers of genes and genomes brings to light a surprise and a puzzle. As we've noted, you've got 3 billion DNA nucleotide base pairs and about 20,000 genes that each encode a distinct protein. Proteins come in a wide range of sizes, but the average number of amino acids in a human protein molecule is about 400, and each of these amino acids is specified by three DNA nucleotides. Therefore, 20,000 distinct proteins require about $20{,}000 \times 400 \times 3 = 24$ million DNA base pairs. The human genome isn't 24 million base pairs long, however; it's 3 *billion* base pairs. The genome is over a hundred times larger than the protein-coding DNA it contains! Historically, we knew the length of the human genome before we knew its sequence of letters and the number of genes it contains; the small number of protein-coding genes compared to what we expected based on the size of the genome came as a shock. For rice the discrepancy is smaller, but it's still around a factor of 10. In general, most of a genome doesn't directly encode proteins. The puzzle, which we're still unraveling, is what the rest of the genome is doing. Some parts are transcribed into RNA but not translated into amino acid chains. These include independently functional RNA segments, as noted earlier, and RNA that is spliced out of the strand transcribed by RNA polymerase before it is translated by the ribosome. Much of the noncoding DNA, however, is never even transcribed into RNA; it can nonetheless influence the readout of genes by making up sites such as promoter regions.

Before expanding on this, let's first develop a better physical picture of DNA. We started this section with the question, "How big is your genome?" and gave a biologically accurate but physically unsatisfying answer: 3 billion base pairs. *How big* is this? Each of the two copies of your genome, housed in nearly all of your cells, would span a meter (3 feet) in length if laid out in a line. The cellular nucleus that contains the DNA is a few micrometers (a few millionths of a meter) in diameter.

Your cells stuff a meter of DNA into a space a thousandth of a thousandth of that length. Is this impressive or not? This may seem like a silly question—obviously, it's impressive. But I can roll 50 yards of yarn into a ball a few inches wide, to the amazement of no one. The central question is one of mechanics: How stiff is DNA? Is it like yarn or like steel? (Hopefully, you *would* be impressed if I stuffed 50 yards of braided steel cable into a few-inch bundle, even if it were the same thickness as yarn.)

DOES DNA BEND?

Characterizing the stiffness of materials is a topic in itself, the exploration of which could take us into detours of materials science that would distract us from biophysics. Thankfully, there's a conceptually simple model of the rigidity of polymers—long, chain-like molecules—that can give us essential insights into genome size. Imagine three different strings of different stiffnesses, each with the same end-to-end length if held taut as a straight line, but left free to form amorphous blobs.

Intuitively, we realize that the one that is most extended, as if made up mainly of long stretches of gentle curves, is the stiffest of the set (left). The most convoluted string, scrunched tight, is likely the softest (right).

Let's think about the typical distance over which the molecule looks straight—in other words, the typical distance we would travel before we start facing a different direction if we're walking along the strand. The stiffer the molecule, the longer this distance. Imagine an ant walking along an uncooked strand of spaghetti—its direction hardly changes at all as it moves along. This characteristic length for the spaghetti is very large. Now imagine a cooked strand of spaghetti, tossed onto a tabletop. Following the strand, our ant's path often bends and turns; this characteristic length is shorter, probably less than an inch or so.

Now let's replace in our minds the molecule's actual curvy path with a series of straight segments, each as long as this typical straight-path distance, connected by joints that randomly orient adjacent segments.

What we've created is something physicists and mathematicians call a *random walk*. Imagine a walker who takes a series of steps in completely random directions—one step might be to the north, the next to the southwest, the next north-by-northeast, and so on. Trying to predict where our random walker ends up after some number of steps sounds futile; who knows where she'll find herself, since the direction of each step is purely up to chance. For any individual walker, prediction is, in fact, futile. But if we imagine taking many random walks, or watching many random walkers, the *average* outcome is well defined: a random walker who takes 25 steps will find herself an average distance of 5 step-lengths from where she started; for 49 steps, 7 step-lengths; for 100 steps, 10 step-lengths; for N steps, the square root of N step-lengths away from the starting point. (This is true whether the walk is two-dimensional, as a person would take, or three-dimensional—perhaps a random swim.)

These random walks turn up in countless places. Economists describe the rapid ups and downs of stock markets as random walks. The paths of swimming bacteria and the spread of random mutations in a population are often modeled as random walks. The list of examples is ever expanding. These trajectories are paradigmatic of our theme of predictable randomness, as robust average properties coexist with the vagaries of chance. In addition, random walks, with their strange dependence of travel distance on number of steps, give us our first glimpse of the general theme of scaling. As we'll learn in part II, many physical characteristics don't simply grow proportionately to size; like our square root above, unexpected dependencies often pop up.

If we think about the DNA configuration abstractly as a random walk, our question of how stiff a DNA molecule is becomes transformed into questions of how long the "step size" is and how many steps there are. From images of DNA molecules, one can deduce that the length over which the double helix "looks" straight is about 100 nanometers, or a tenth of a millionth of a meter. In other words, replacing the actual path of the DNA with straight lines, and asking what straight-line length is appropriate, gives a value of 100 nanometers. (The technical name for a polymer's straight-line length is the Kuhn length, after Swiss chemist Werner Kuhn, and there is a precise mathematical expression for its calculation.) For scale, the width of the double helix is 2 nanometers, and the length along the ladder for a full helical turn is about 3 nanometers; the Kuhn length is large compared to the fine structure of the helix.

We can think of 1 meter of DNA as therefore being composed of 10 million straight-line steps. Our question of how big 1 meter of DNA would be if left alone, floating around in the watery environment of a

cell, becomes transformed into the question of how far a random walk of 10 million steps, each 100 nanometers long, would be. The answer: about 0.3 millimeters, or 300 micrometers. (That's the square root of 10 million, or about 3000 step-lengths, multiplied by the step-length of 100 nanometers.) That's far larger than the few-micrometer size of the cell nucleus; it's much larger even than the entire 10–100 micrometer width of a typical human cell.

You might object, perhaps, if you're aware that your DNA isn't one unbroken strand but rather is divided into 23 chromosomes. (Nearly all of your cells have 46 fragments grouped in pairs, from each of the two copies of your genome. Exceptions are egg and sperm cells, which have one copy of your genome, and red blood cells, which in humans and other mammals lack DNA.) The fragmentation simplifies the spatial challenge of packaging, but not by much: human chromosome 1, the largest chromosome, is 249 million nucleotide base pairs in length, corresponding to an overall length of about 8.5 cm and a random walk "blob size" of about 90 micrometers, which is still much larger than the size of a cell nucleus. To scale, I've illustrated below a typical human cell and its nucleus, the random blob configuration of 1 meter of DNA, and the random blob configuration of 8.5 cm of DNA (like chromosome number 1).

We should, in fact, be amazed and impressed by the packing of DNA—not because of the length of your genome, but because DNA is so stiff that it isn't amenable to being confined inside a cell. The space

cell

chromosome 1

1 meter
DNA

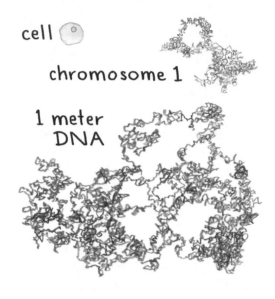

into which it's packed is far, far smaller than the space the molecule would occupy if left alone, floating in a watery world.

DNA PACKAGING

The DNA inside our cell nuclei isn't left alone like our spaghetti strands or random walks, and isn't stuffed in like clothes in a hastily packed suitcase, but rather is elegantly and compactly packaged. Much of your DNA is wrapped around little spools about 10 nanometers in diameter, made of proteins called *histones*.

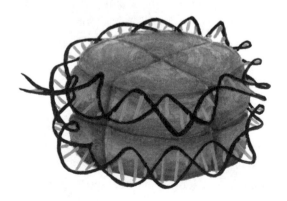

Ten nanometers is much smaller than the Kuhn length of DNA, so this wrapping requires a lot of force, provided in large part by electrical attraction between the DNA, which is negatively charged, and the positively charged outer surface of the histones. The spacing of positive amino acids matches the periodicity of the double-helical grooves, maximizing the magnitude of electrical forces. Once again, we find self-assembly at work: the physical attributes of DNA and histone proteins, especially their charge and shape, enable them to craft themselves into a well-defined, functional structure. Nearly two turns of DNA, or about 150 nucleotide base pairs, are wound around each spool. The span in between the spools varies in length, between 20 and 90 base pairs, and the whole assembly is evocatively referred to as "beads on a string."

These strings of spooled DNA are further looped and packed together. What form they take has been a long-standing mystery, with a variety of structures proposed based on experiments that typically involve extracting DNA from cells or preserving cells with fixatives. The most common picture is of the beads on a string organizing into a 30-nanometer-thick fiber, with these fibers then arranged into 120-nanometer and thicker cords. Very recently, however, researchers at the Salk Institute and the University of California at San Diego, led by Clodagh O'Shea, developed a method to stain DNA in intact nuclei, decorating them with metal atoms that are readily visible in an electron microscope. Applying this approach, they did not find the expected discrete fibers, but rather uncovered chains with a broad spectrum of widths, ranging between 5 and 24 nanometers. These chains, moreover, differed in how curly or straight they were depending on whether or not cells were dividing. It may be that the packaging of DNA is less stereotypical, and more dynamic, than was previously thought.

The means by which cells stuff DNA inside themselves is more than an intellectual curiosity. The expression of genes—whether some span of DNA is actually transcribed into RNA, allowing the creation of proteins—depends a lot on the packing and organization of DNA. Regions of DNA that are wound around the histone spools or that are otherwise tightly constrained are relatively inaccessible to the machinery that reads and executes the genetic code. The exact same gene can be "on" or "off," depending on whether it is easily found or hidden away. DNA packaging, in other words, affects DNA function, and the physical arrangement of DNA is a powerful tool for regulating the activities of the cell. A variety of maladies as diverse as neurodevelopmental disorders, rare autoimmune diseases, and even cleft palate are associated with flaws not in the genes that encode the proteins that perform the associated neurodevelopmental, immunological, or skeletal tasks but rather in the DNA packaging. These flaws often involve the proteins that manipulate histone spools, increasing or decreasing their affinity for each other or for DNA, for example, by changing their charge.

In the past two decades or so, scientists have discovered that the determinants of what DNA regions are wrapped around histones are embedded in the DNA sequence itself, dictated in part by the mechanical properties of the double helix. We saw earlier that DNA is quite straight over lengths of about 100 nanometers. The exact stiffness depends subtly on the DNA sequence (the As, Cs, Gs, and Ts). Particular groups of nucleotides are less stiff than others, or by virtue of their shape prefer a slight curvature. In DNA that ends up wound around nucleosomes, these more curved or flexible regions, like little hinges, tend to be situated 10 nucleotides apart. The pitch of the double helix is also 10 nucleotides, meaning that if you stood on the twisted ladder of DNA and climbed 10 rungs, you'd be facing the same direction you were to begin with. The hinges, therefore, are all oriented the same way, allowing each span of DNA to bend toward the histone spool. Analysis of the binding between DNA and nucleosomes shows that sequences without these repeating pairs of nucleotides are less likely to be wrapped around histones. The DNA sequence itself, therefore, encodes mechanical information about how it should be packaged. DNA is a mechanical code subtly interleaved with a biochemical and genetic code—truly an extraordinary molecule!

The architecture of spools and fibers of DNA provides an example of the general theme of regulatory circuits with which cells can control their activity, turning genes on or off. As we'll see in chapter 4, many more strategies exist, enabling faster and more complex decision-making circuits.

DNA-STUFFED VIRUSES

It's not just *your* cells that are faced with the task of compressing DNA into tight spaces. Every living organism packs its DNA, and none leave it alone as a free, random-walk chain. This even extends to the not-quite-living world: viruses, little capsules of genetic material that infect living cells, hijacking their replication machinery, contain the densest-known packaging of DNA. Not all viruses contain double-stranded DNA; some contain a single DNA strand and some contain single- or double-stranded RNA. Those that do have a double-stranded DNA genome, which include the viruses that cause herpes and smallpox, must stuff this rigid molecule into a protein shell just tens of nanometers in diameter, again smaller than the Kuhn length of the DNA double helix. (Double-stranded RNA is even stiffer than double-stranded DNA. For both DNA and RNA, a single strand is much more flexible.)

In a double-stranded DNA virus, the bent, squished polymer pushes on the virus's shell, or *capsid*, trying to stretch out. When the capsid is opened, for example, when the virus infects a cell, this internal pressure helps propel the DNA into its cellular target. How can we measure the pressure of the compressed DNA? Imagine opening a closed capsid; the DNA rushes out. Now imagine squeezing the capsid from all sides, applying pressure, and *then* opening the capsid. If the external

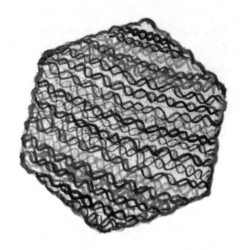

pressure is less than the internal pressure, DNA will still come out. If the external pressure is greater, the DNA will remain inside. By varying the external pressure and monitoring whether or not DNA is released, one can determine the pressure inside the virus.

That's easy to imagine, but actually doing it requires coming up with some clever experimental tricks, one of which was implemented about 15 years ago by William Gelbart and colleagues at the University of California at Los Angeles. Capsid opening is naturally triggered when a virus encounters particular proteins on the surface of its target cell. Adding these proteins artificially to a beaker full of viral capsids provides opening on demand. The viral particles are dispersed in a watery solution. Adding large molecules to the solution provides an *osmotic pressure*—a bit like bombarding the virus with all the molecules floating around it—that acts like our hypothetical squeezing of the capsids. By varying osmotic pressure and using protein-triggered capsid opening, scientists discovered internal pressures in viruses that were tens of atmospheres in magnitude. (For comparison, the air pressure in a car tire is about two atmospheres.) To get a more intuitive sense of the mechanical feats performed by these viruses, biophysicist Rob Phillips suggests envisioning 500 yards of Golden Gate Bridge suspension cable crammed into the back of a FedEx delivery truck. These huge internal pressures are valuable for the virus, helping it launch its DNA into a targeted cell where it will be replicated, initiating the generation of new viruses.

We can't understand DNA without understanding its physical properties. Shape, structure, and mechanics are inextricably tied to biological function. This statement isn't true just for DNA but for all of nature's biomolecules—a recurring theme throughout biophysics. In the next chapter, we return to the question of how a surprisingly small number of genes can guide the processes that make *you* by exploring how genes can be switched on and off—by external controls or by other genes—creating a meshwork of interactions that is again inseparable from the tangible, physical activities of life's molecules.

The Choreography of Genes

Chapter 3 introduced the central puzzle of genetic information: How is it that a mere 20,000 genes encodes the complexity of *you*? How do just 20,000 proteins—in other words, 20,000 tools or 20,000 components—perform the dazzling array of tasks that you're capable of, from growing to breathing to reading to reproducing? This is, of course, a human-centered way of asking these questions—we could just as well ask how 20,000 genes make a horse a horse, or how 30,000 genes make a water flea a water flea.

We're far from having a complete answer to these questions, and their exploration will keep scientists busy for decades, or centuries, to come. We have, however, uncovered intriguing general principles, motifs that illustrate how the complexity of life is encoded, and we've begun to use these principles to engineer living organisms in unprecedented ways. In the previous chapter, we viewed genes as more or less static—packaged in the space of the cell, with the potential to dictate the assembly of proteins. Now we introduce time—stimulating or suppressing the transformation of genetic information into physical activity as needed by dynamic, living organisms. Much of this choreography of genes is organized by genes themselves. We've so far considered self-assembly in a concrete, structural sense. Here we encounter a more abstract manifestation of self-assembly, as molecular activities weave themselves into regulatory circuits that make every creature a biological computer. To see what this means, we start with the notion of turning genes on and off.

GENE REGULATION

A cell, or a whole organism, can control when and whether to actually make use of any given gene—in other words, whether its string of As, Cs, Gs, and Ts is read by the machinery that transforms a DNA sequence into an RNA sequence into a protein. This control can be influenced by the external conditions that the cell or organism is experiencing, letting it activate or deactivate particular genes in response. Even before understanding regulation in detail, you could infer that something like this must exist from the fact that your body is composed of very different types of cells, though each contains a copy of the same DNA. The genomes inside a neuron, a skin cell, and a mucus-secreting cell that lines your gut are all identical. These cells, however, don't look the same, don't perform the same activities, and are not synthesizing the same set of proteins. Genes for proteins that create mucus must be dormant in neurons; genes for proteins that adhere tightly to neighboring cells must be active in your skin; your secretory cells must ignore the genes responsible for sending long-distance electrical signals. Somehow, it must be possible to turn genes "on" and "off." Let's see how this control is made possible.

Recall the transcription of a DNA sequence into RNA. The RNA polymerase machine slides along DNA like a train on a track, transcribing the sequence of a gene from its start to its stop signals, extruding a strand of RNA as it glides along. RNA polymerase isn't always bound to DNA, however. Much of the time it floats around in the watery medium of the cell; occasionally, it bumps into a specific DNA sequence that it recognizes and latches onto. As we saw in chapter 3, these sequences, called promoter sequences, are adjacent to genes or sets of genes. There's a directionality to DNA, and an RNA polymerase "reading" one strand of the DNA double helix will travel in a particular direction. Genes lie downstream of their promoters, so that an RNA polymerase that lands on a promoter sequence transcribes the adjacent genes. Controlling the tethering of RNA polymerase is the essence of *transcriptional regulation* of genes, one of the most powerful ways to control gene activity.

We first figured out mechanisms of transcriptional regulation in bacteria. Imagine you're a bacterium. You like to eat sugars, but you need to make sugar-digesting proteins to do this. You'd prefer to make more such proteins if you encounter sugar, and you'd rather not waste energy making these proteins if there's no sugar around. How can you do this? We illustrate with an actual sugar, lactose, and the regulatory machinery found in the bacterium *E. coli*, which is very similar to machineries used throughout the living world.

A gene called *lacZ* encodes part of the lactose consumption machinery. Upstream of it, as always, is its promoter region. I've drawn RNA polymerase as a green blob poised to advance and read the (blue) *lacZ* gene.

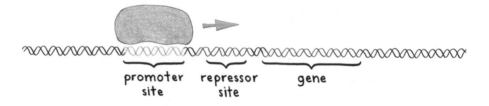

(The illustration isn't to scale; the *lacZ* gene is about 3000 base pairs long, for example, and the width of RNA polymerase spans 30 to 40 base pairs.) *E. coli* also makes a protein called the lac repressor, which binds to another stretch of DNA upstream of the *lacZ* gene. When the lac repressor (red) is DNA-bound, RNA polymerase can't attach and the *lacZ* gene is not expressed.

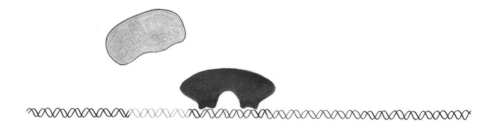

As we've seen in previous chapters, DNA and proteins are physical objects, with particular structures that guide how they work. The binding between the lac repressor and DNA has a particularly striking arrangement. The specific sets of nucleotides that the lac repressor recognizes are spaced farther apart than the repressor's width. The protein must therefore loop the DNA into a tight circle, about 10 nanometers in diameter.

Recall, however, that DNA is very stiff. Left alone, it will be quite straight over distances of about 100 nanometers. Like a circus strongman flexing an iron bar, the lac repressor bends the DNA. The looped DNA interferes with the normal binding of RNA polymerase, preventing the readout of the genes for lactose-digesting proteins.

The lac repressor has another amazing property: it can also bind to a lactose lookalike called *allolactose* (the black circle in the illustration below); when allolactose-bound, the shape of the repressor subtly changes so that it can no longer bind to DNA. The bacteria create allolactose from lactose itself. If the bacterium encounters lactose in its environment and internalizes some of it, so that the lac repressor no longer represses, the lactose-digesting proteins are made and the bacterium can gorge on the food it has found.

Repressors like the lac repressor are common in all organisms, not just bacteria. Inhibiting RNA polymerase binding, or at least competing with it to make it less likely, is one of nature's favorite tactics

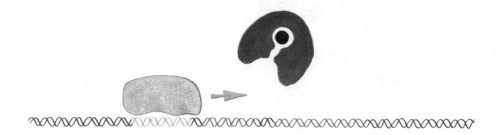

for regulating the activity of genes. As with the lac repressor and lactose, this inhibition can be coupled to an external stimulus or, as we'll see, an internal stimulus.

The opposite of repression in the tool kit of genes is activation. Especially for promoters to which the binding of RNA polymerase is weak, activator proteins with an affinity for RNA polymerase can attach to nearby DNA, again at specific sites recognized by the proteins, enhancing the likelihood that RNA polymerase will stick and initiate transcription.

Activator proteins also have a role to play in the story of lactose and bacteria. Though bacteria such as *E. coli* like lactose, they like glucose, another sugar, much more. If glucose is present, they won't waste their efforts digesting lactose even if it's available. This phenomenon was discovered in the 1940s by Jacques Monod, who balanced studies of fundamental biology with work in the French Resistance during the Second World War. At the level of its DNA, the bacterium must express the lactose-digesting genes only if lactose is present *and* glucose is absent. An activator protein makes this possible. The binding between RNA polymerase and the lac promoter region is weak, so even if the repressor is absent, the transcription of the gene is unlikely to occur. The bacterium makes an activator protein that binds to DNA only if it has also bound a molecule called *cyclic AMP*. Cyclic AMP is produced by

the bacterium when glucose levels are low; it's sometimes called a "hunger signal." Therefore, if glucose is present, there's little cyclic AMP, the activator won't bind, and the lactose-digesting gene won't be expressed even if lactose is present. If glucose is absent, there's a lot of cyclic AMP, the activator binds, and the lactose-digesting gene is expressed *if* the polymerase isn't being blocked by the lac repressor. It's a clever system, especially for a brainless creature a thousandth of a millimeter in size.

Repressors and activators are both referred to as *transcription factors*—things that govern the transcription of genetic information. Transcription factors are themselves proteins, and so are encoded by genes. We have a lot of them—it's not known exactly how many, but it's thought that the human genome contains over 1500 genes for transcription factors. Recall that we have only about 20,000 protein-coding genes. A sizable fraction of our genetic instruction set, in other words, is made up of the brakes and levers that regulate the readout of those instructions.

Transcription factors and the decisions they make possible are found throughout the living world and are essential for the encoding of complex behaviors by simple genes. Regulatory regions—the landing sites on the genome for transcription factors—don't even need to be adjacent to the genes they regulate. Because the genome twists and turns, a transcription factor bound to a segment of DNA can influence expression of a gene that would be distant if the DNA were laid out in a straight line but that's nearby in actuality.

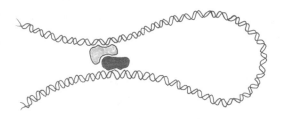

This interplay between genetic and physical distance offers even more possibilities for gene regulation and is the topic of a great deal of present-day biophysical investigation.

All the mechanisms we've explored so far involve the regulation of transcription, the first step in gene expression in which a DNA code is turned into an RNA code. Cells can also regulate *translation*, the synthesis of a protein from the RNA segment. This too can be done in many different ways, including controlling the rate at which RNA degrades, cloistering RNA in particular regions of the cell, and even creating RNA molecules that are complementary to the transcribed RNA so that the two bind together to form a double-stranded RNA that can't be translated into protein. We could spend many more pages exploring the variety of the gene regulation tool kit, but instead, let's step back and look at the universality of these tools and some motifs nature has developed to combine these tools into machines.

PORTABLE GENETIC CONTROL

We've seen how the lac system uses transcription factors to turn a gene on or off depending on stimuli from its environment, namely, the presence or absence of certain sugars. *E. coli* and other bacteria use this system to match their biochemical activity to the availability of particular foods. A researcher could easily add lactose to a flask full of glucose-starved bacteria, triggering the microbes to activate the *lacZ* gene. More sneakily, though, the researcher could add a chemical called IPTG, which is very similar to lactose except that it can't be digested. The lac repressor will bind to IPTG and therefore not repress transcription, and the cell will produce lactose-digesting enzymes even though there's no lactose to digest. The motivation for this strange subterfuge is to construct a handle for gene expression. The researcher beforehand could have inserted other genes downstream of the *lacZ* promoter, perhaps also deleting *lacZ* itself. These new genes could be genes for fluorescent proteins with which to monitor the bacterium, or genes to synthesize various useful chemicals or drugs. The researcher now has the expression of these new genes under the control of an external trigger, the IPTG.

A striking example of this genetic control is described in a paper from 2001 by Heidi Scrable and colleagues at the University of Virginia, simply titled "The Lac Operator-Repressor System Is Functional in the

Mouse." The researchers made use of albino mice, which have a mutation in a gene called tyrosinase that is necessary for pigment production. By inserting a functional tyrosinase gene (brown in the illustration) into the genome, along with its promoter sequence (green), the
authors created mice with normal pigmentation: brown hair and eyes,
as one would expect. The remarkable part comes from engineering control of the pigmentation genes. Though animals have a large number
of gene regulatory systems, the lac system is not one of them; it's only
found in bacteria. However, the researchers engineered a mouse that
contains the DNA binding sequence of the lac repressor (red), situated
upstream of the functional tyrosinase gene. Because the lac repressor
gene doesn't exist in mammals, there's no lac repressor protein, and
no repression of tyrosinase. These mice, therefore, are also pigmented
(second row).

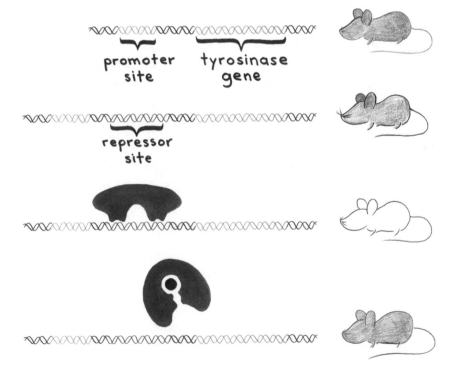

Yet another set of mice was created in which, again, tyrosinase was controlled by the lac promoter but in which the lac repressor gene was
also inserted into the genome, along with its promoter, at some other

location. These mice produce the lac repressor protein, which represses tyrosinase expression, so the mice lacked pigmentation (third row). If the researchers added IPTG to the drinking water of the mice, the animals produced the proper pigmentation, turning brown (fourth row). The IPTG prevented the lac repressor from repressing, just as it does for bacteria, enabling the expression of the pigmentation gene.

In addition to the almost surreal ability to change an animal's hair and eye color by adding a sugar-like molecule to its water, this outcome highlights the universality of life's machinery. The last common ancestor of mice and bacteria lived over three billion years ago. The descendants of that ancestor have evolved along separate paths ever since, giving us very different creatures—a single-celled microorganism and a palm-sized, furry mammal. Nonetheless, one can cut-and-paste a regulatory apparatus from one into the other, and it works perfectly well. As Monod himself dramatically and presciently noted decades earlier, "What is true for *E. coli* is true for the elephants."

Like the lac system, there are many others that allow organisms, or researchers, to regulate gene expression. My own lab also makes use of these sorts of engineered circuits. We don't change the colors of mouse fur, but we turn on or off the ability of certain bacteria to swim, directing them to assemble or disassemble their microscopic motors by placing a simple chemical into their water. With this tool, we can infer the role of a physical behavior like swimming on the bacteria's ability to succeed in their environment. Just a few decades ago this would have been science fiction; now it's not only possible but becoming steadily easier.

GENETIC MEMORIES

If you flip the switch to turn on a lamp, you needn't keep pressing your finger against the button for the light to remain on. The switch is toggled to a new, stable position, and it stays there until toggled to a different, also stable, position. Nature, and researchers, also often want switches like this, which set cells on a particular path once they've received a signal and keep them on that path even after the signal has gone. In plants and animals, this is especially important for the development of different types of cells. Both neurons and the glia that help neurons function, for example, descend from the same type of progenitor cell. Some signal sends a progenitor on the path to be a neuron, after which it's committed to express the proper set of neural genes, make synapses with other neurons, and perform all the other tasks a neuron is charged with. You wouldn't want it to need constant reminders not to revert to its ancestral form, or to suddenly switch to being a glia, or to turn into a confused half-neuron, half-glia. Having stable cell types requires having toggle switches for genes. Another way of saying this is that the cell needs *memory*; it needs to remember past stimuli, encoding them into the way genes are expressed in the present and future.

There are many ways to make a memory. Here's one that builds on what we've seen of transcription factors. Imagine two genes that we'll call A and B. As with lac, imagine there's a repressor for gene A. Now suppose that the gene that encodes that repressor protein is immediately downstream of gene B, so that if B is expressed, the A-repressor is also expressed. Now, symmetrically, imagine that the B-repressor gene is just downstream of A, so that if A is expressed, the B-repressor is also expressed. This mutual repression enables memory: Suppose A happens to be strongly expressed. The cell makes lots of B-repressor protein, so B will be repressed and there won't be any repression of A, consistent with A's strong expression. The cell will continue in the A state. On the other hand, if B is strongly expressed, the opposite set of events will occur, and the cell will continue in the B state. We have two stable types of behavior for the cell. We can toggle between them by, for example, flooding the cell with a lot of whatever triggers the activation or repression of A or B—IPTG, for example, if the lac repressor is used for part of this apparatus. From that point on, the cell retains a memory of this event.

There's a very general principle illustrated here, which is that genes regulate the expression of genes. In other words, feedback between genes creates particular patterns of activity. Our example uses two instances of repression (negative feedback) to create a switch. It's not just hypothetical: exactly this scheme is used throughout nature, for example, in viruses infecting bacteria that decide between an actively dividing and a dormant state. Many other schemes are also possible. We could use activators of transcription, with, for example, expression of gene A coupled to expression of an activator for gene A, amplifying whatever initially put the cell on the A path (positive feedback).

CLOCKS AND CIRCUITS

To tell time, we use clocks. Every clock is based on some periodic, rhythmic phenomenon, such as the back-and-forth swaying of a pendulum bob or the rapid vibrations of a quartz crystal. All sorts of living organisms, and even single cells, use clocks to control activities that should wax and wane with some well-defined period. The circadian rhythm is a great example. In many plants, the production of chlorophyll has an approximately 24-hour cycle, matching the periodicity of the day. The plant doesn't solely rely on external cues, subject to the whims of clouds and shadows, but has an internal timekeeper with a 24-hour period. You do as well, with body temperature, blood pressure, and of course sleepiness rising and falling roughly once per 24 hours, even if you're sequestered in a windowless, constantly lit room for weeks on end. Many animals, fungi, and even some bacteria possess a circadian clock. Sensing light helps maintain the rhythm and shifts the timings of its peaks and troughs, but the periodicity itself arises from internal oscillators.

Gene regulation provides a way for a single cell to make an oscillator, using only the ingredients of the cells themselves. We'll be a bit abstract, since the details of real cellular oscillators are rather complex, involving many interacting genes. The basic idea, though, can be illustrated with just one gene.

The simplest possible oscillator consists of a gene that represses itself—in other words, a gene that encodes a protein that represses the expression of that gene itself. At first glance, this seems ridiculous; how could such a gene ever be expressed? The answer lies in the fact that both expression and repression take time. Recall that the expression of a gene means the transcription of that stretch of DNA into a segment of RNA and then, afterward, the translation of that RNA into a chain of amino acids—a protein. After that, if the protein is to repress the gene, it must meander and find the promoter region. All this takes time. Even after the repressor binds and RNA polymerase is blocked, the activity of the gene isn't immediately quenched. The pieces of RNA that were already made can continue to be translated into protein, and the proteins that were already made can continue to do whatever they were doing. The upshot of all this is that the expression of the gene can increase for some time. The activity of the gene, therefore, persists for a while, though it represses itself. To see how this activity can oscillate, we need one other fact about proteins: they all degrade over time.

The decay of transcription factors has a large impact on their regulation of genes. A fundamental truth of the physics of molecular binding is that the greater the concentration of a molecule, like our repressor, the greater the likelihood that it will be bound to something it has an affinity for, like our promoter region. Conversely, as the free repressor proteins degrade and their concentration declines, the repressor proteins on the DNA become increasingly likely to unbind, no longer repressing transcription. The gene can then be expressed.

Putting this all together, we have the following: In the preceding circuit, the gene is initially expressed and slowly builds up the concentration of its own repressor proteins, inhibiting further creation of proteins. But the existing proteins degrade, and eventually so many have vanished that the gene can be expressed again, and the cycle begins anew. Thus, we can have an oscillator.

It's not a very good oscillator, though, and I don't know of any organism that actually uses a clock made of a single gene. The timing resulting from this arrangement is hard to tune and its periodicity isn't very precise. Both properties depend on the rate of protein degradation in a cell, which is determined by a variety of factors beyond the purview of the gene and its self-repression.

A better scheme involves three sets of genes, A, B, and C, in which A represses B, B represses C, and C represses A. I won't give a detailed analysis, but this circuit also oscillates. The amount of each of the A,

B, and C proteins goes up and down periodically. The frequency of the oscillations depends on the affinities of the repressor proteins to the DNA. We, or the cells, can tune the periodicity of the cycles by using repressors with stronger or weaker DNA binding. This A-B-C circuit, called the *repressilator*, doesn't exist in nature, at least not as a stand-alone unit. However, it was one of the first to be artificially engineered into cells, by biophysicists Michael Elowitz and Stanislas Leibler at the turn of the twenty-first century. The researchers inserted this oscillator into the genome of *Escherichia coli* bacteria and coupled it to the green fluorescent protein gene, creating cells that rhythmically switched between being fluorescent and dark. Since then, a variety of other precise and tunable oscillators have been engineered into cells.

Though a stand-alone repressilator hasn't been found in nature, similar circuits are commonplace. The 24-hour oscillator that controls the human circadian rhythm, for example, involves a handful of genes connected by intertwined feedback loops, including the repressilator motif. The resulting machinery gives rise to a clock that is robust but also trainable by external stimuli, such as sunlight, that induce various chemical changes in our bodies. This training isn't immediate, as anyone who has experienced jet lag knows. Sometimes, we'd like to reset our clock more rapidly than our bodies can accommodate. Our circadian rhythm, however, didn't evolve in a world with high-speed air travel.

Beyond memories and oscillations, there are innumerable other combinations of gene interactions made possible by the regulatory tool kit. Imagine, again abstractly, an A-B-C trio of genes in which A and B each encode an activator of C, and the expression of C is weak unless one of the activators is present. If whatever induces expression of A is present, C will be expressed, *or* if whatever induces expression of B is present, C will be expressed. We could also design some set of genes that only activate C if the A stimulus *and* the B stimulus are present, or if the A stimulus *and not* the B stimulus is present, and so on. (In fact, we've seen earlier in this chapter an example of the last configuration, in bacteria in which *lacZ* is expressed if lactose *and not* glucose is present.)

A device that can perform logical calculations on inputs—decisions based on *and, or, not,* and other such operations, and combinations of these operations—is a computer. The computers we're used to thinking about use electrical voltages—high or low, on or off—rather than the presence or absence of biochemical transcription factors, but the conceptual framework is the same. Moreover, the underlying generality is the same. With the appropriate combination of logic elements, whether they're electrical bits or genes, one can perform any computation, whether it's compressing a digital video file or deciding whether conditions warrant germination of a seed.

With logic and memory, nature can make all sorts of genetic circuits that perform all sorts of tasks. The complexity of gene-driven activities, therefore, can be vastly greater than one might think from simply tallying the number of genes.

GENES IN THE ATTIC

Another way of controlling whether genes are expressed or not involves a phenomenon we've already encountered: the packaging of DNA. As we've seen, the arrangement of histone-wrapped DNA can dictate how easily RNA polymerase can read a gene. Over the past few decades, we've learned how powerful this packaging-based gene regulation is. A tool kit of proteins adds and subtracts particular sets of atoms from histone proteins, influencing their further assembly or disassembly as dense DNA-histone fibers. This modification of histones is particularly important during the early development of an embryo, as descendants of a handful of cells permanently adopt distinct cellular identities, and in cancer, as cells change into rapidly growing and migrating forms. Genes that aren't needed by particular cell types remain packed away, as if in storage in an attic—available, but not readily accessible.

Cells can also alter DNA itself to affect its readability. Certain proteins, for example, can perform chemical reactions that replace a hydrogen atom with a carbon atom and three hydrogen atoms (a methyl group) on an A or C nucleotide. This appendage can inhibit the expression of the gene the methylated nucleotide is part of. This tiny tag on the DNA can gum up the transcription machinery and can also recruit proteins that further modify histones.

Surprisingly, it seems that these facets of genetic regulation—packaging and methylation—can be passed on from parent to child. We inherit, it seems, not just our genome from our parents—the sequence of As, Cs, Ts, and Gs—but some aspects of the genome's organization that affect how it is interpreted. This phenomenon, known as *epigenetics*, adds further complexity to the connections between our genetic code, our environment, and how we function as organisms. Studies of the Dutch who endured the 1944–1945 "hunger winter," for example, found an increased likelihood of obesity and cardiovascular disease later in life, and altered DNA methylation that persisted for decades; moreover, similar signatures of ill health were present in the survivors' children, born long after the famine, indicative of epigenetic inheritance.

We conclude this chapter by reminding ourselves that the machinery of gene regulation makes every organism a powerful, versatile computer, capable of making decisions based on the diverse stimuli provided by its environment, and orchestrating behaviors that vary in time and space. This complexity doesn't require thought, or central control, but rather comes from the nature of the genetic code itself, which contains genes as well as the means to regulate them. Again, we see self-assembly at work, with machineries that construct themselves. The elegance of genetic circuitry and predictable decision making coexists, however, with an inherent randomness that is ubiquitous in the microscopic world, the importance of which we see in chapter 6. First, however, there's one more crucial cellular component to introduce: membranes, whose architecture provides a stunning example of self-assembly separate from DNA and proteins.

Membranes: A Liquid Skin

You began as a single cell, a fertilized egg. This cell divided in two, these two into four, and after many more divisions, rearrangements, and changes of shape, you developed a body composed of tens of trillions of cells. Every one of these cells has a membrane at its edge. This membrane doesn't just serve as a boundary between inside and outside, but forms an arena where the cell grabs onto its surroundings, traffics chemicals, and exchanges signals with its neighbors.

Membranes exist within cells as well, as the boundaries of a variety of organelles ("little organs"). The nucleus that houses each of your cells' DNA, for example, is an organelle (purple in the illustration). Many proteins are made at an organelle (the endoplasmic reticulum, dark blue) that consists of a long, convoluted labyrinth of membrane. At another organelle (mitochondria, red), the cell synthesizes the chemicals it uses to transport energy; this organelle has a double membrane, the inner of which is folded into layered stacks. Not all cells have

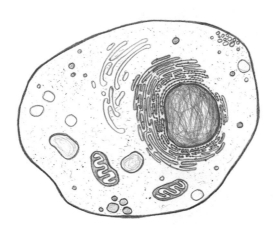

membrane-bound organelles: members of the bacteria and the archaea, two of the three domains into which all creatures are classified, lack them. Organelles, however, are prevalent in the third domain, that of the eukaryotes, which includes all animals, plants, fungi, and many single-celled organisms, such as amoebas.

At the core of every cellular membrane is a sheet just a few billionths of a meter thick made of molecules called *lipids*. Proteins are present as well, poking through the membrane, opening pores in it, or lying against it. These membrane-associated proteins account for over a third of the human genome and form the majority of pharmaceutical targets; they orchestrate a lot of biological activity. The lipids, however, are what make a membrane a membrane. Given the importance of membranes, one might expect their structure to be precisely dictated by the cell's genetic code and carefully monitored by some internal machinery. Instead, we find the opposite, laissez-faire approach: proteins make lipids and self-assembly does the rest. Physical interactions among lipids and water suffice to generate a reliable yet dynamic material. To understand the properties of membranes and how they emerge, let's first look at something more familiar.

HOW TO MAKE A MEMBRANE

Oil and water don't mix. The oil in a shaken bottle of vinaigrette salad dressing quickly coalesces into droplets. Left alone, oil molecules associate with other oil molecules and water molecules with water molecules. As we saw in chapter 2, substances such as oils and fats that separate from water are called hydrophobic ("water-fearing"); substances such as sugar and vinegar that mix with water are hydrophilic ("water-loving").

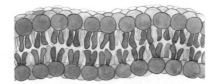

Lipids are both hydrophobic and hydrophilic. Each lipid molecule has a "head" that likes water and a "tail" that doesn't. The tail typically consists of two chains, each of which chemically resembles an oil molecule. There are other familiar substances that also have these amphiphilic ("loving both") tendencies: every soap molecule also has a hydrophilic head and a hydrophobic tail, typically just one chain, which together let the soap cling to greasy dirt as well as to the water that washes it away.

Lipids in water struggle with the contradiction imposed by their structure: their heads are happy but their tails are not. They therefore spontaneously associate with one another to shield their tails from the water, forming a two-molecule-thick sheet. This "lipid bilayer" forms the basis of all cellular membranes.

Lipid membranes have remarkable physical properties. They are essentially two-dimensional: the bilayer is about 5 nanometers thick—about 20,000 times thinner than a typical sheet of paper—while its lateral extent can be many thousands of times greater than its thickness. They are flexible, bending and curving in three dimensions, and living cells carefully control this curvature.

You might think that a membrane is rather like a plastic bag: thin and flexible. There's a key difference, though. If you take a marker and dab a spot of ink on a plastic bag, then leave it alone and check back a few minutes later, the spot will be in the same place you left it. If you were to do the same to a lipid membrane, the spot would blur and disappear, the marked molecules having meandered elsewhere throughout the membrane. Lipid bilayers and cellular membranes are fluids. Just as the water molecules in liquid water aren't fixed in relation to one another, but rather can flow throughout the fluid, lipids and proteins embedded in a membrane are mobile and can flow throughout its extent. Membranes are two-dimensional fluids. This mobility, like the other physical properties so important to membranes, follows from the nature of lipid bilayers: the lipids aren't rigidly bound to one another, but are simply associating to shield their hydrophobic tails from water. There's no impediment to molecules weaving around one another, as long as the hydrophobic core of the bilayer remains unexposed to water.

Molecular mobility is a wonderful trait: membrane molecules can rearrange themselves, interact with one another, and even form structures and patterns that help cells carry out various tasks.

A striking example occurs in your immune system, during the interactions between two cell types, known as T-cells and antigen presenting cells (APCs). APCs take up proteins from their surroundings, rip them apart, and put the resulting fragments on display by attaching them to proteins sticking out from their outer membranes. T-cells meet the APCs, make contact, and examine the fragments to determine whether they came from you or from some foreign source—perhaps a bacterium or virus. If foreign, the T-cells activate your immune system, triggering your body's defenses to fight the apparent invasion. This response to foreign protein fragments involves a dynamic molecular dance at the T-cell/APC contact. Adhesion proteins at each cell's outer membrane bind to one another and start to cluster together.

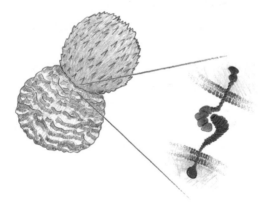

Around them, the proteins involved in the display and recognition of protein fragments—in other words, the signaling between the cells— also begin to bind to one another and group together. If we imagine the plane of this page as the interface of contact between the two cells, the initial arrangement, with adhesion and signaling proteins drawn as blue and red, respectively, looks like the illustration on the left.

Then, over the course of a few minutes, this bulls-eye inverts, and hundreds of signaling proteins flow to the center while the adhesion proteins assemble in a ring (right).

This structure, called the *immunological synapse,* was discovered in the mid-1990s. Understanding how these spatial patterns form, and then how a T-cell translates them into the "on" signal for its activation, has been intensely studied in the years since. In addition, scientists have found that similar synapses form at the contacts between immune cells transmitting the human T-cell leukemia virus as well as the human immunodeficiency virus (HIV, which causes AIDS). These viruses, it seems, have figured out how to hijack the structure-forming machinery of cells they infect. We won't go into the intricacies of synapse formation but limit ourselves to pointing out that, if they weren't embedded in a two-dimensional fluid, T-cell signaling proteins, adhesion proteins, and many other membrane proteins in many types of cells would be incapable of performing the dynamic spatial rearrangements that nature demands of them.

Protein and lipid patterns also intersect the theme of predictable randomness. The fluidity of membranes allows their components to reorganize, but it brings with it the uncertainties of flow and disorder. Neither we nor a cell can know exactly where every lipid and every protein will be, but we can predict, on average, properties of the ensemble. Both the deeper nature of the randomness and the meaning of prediction become clearer in chapter 6.

CONES, SPHERES, AND BUBBLES

A bilayer isn't the only structure that an amphiphilic molecule can form. Consider a molecule shaped like an ice cream cone, with the hydrophilic part being the ice cream and the hydrophobic part being the cone (left). In water, it will self-assemble into a sphere (right), illustrated two-dimensionally here, to shield the hydrophobic cones.

Molecular shape is a key determinant of self-assembly. Most of the lipids in your body are "cylindrical" rather than conical; the hydrophilic heads and hydrophobic tails are similar in width, hence the ensemble of lipids forms into a fairly flat bilayer. A small fraction of your cells' lipids are not cylindrical, though—not enough to disturb bilayer formation, but enough, it's thought, to help generate curvature as the cell bends membranes into complex structures.

You might wonder whether amphiphiles could be arranged so that the hydrophilic heads are "in" and the hydrophobic tails are "out." They can, and you've created such a structure every time you've blown a soap bubble.

At the edge of a soap bubble, soap molecules sandwich a thin film of water, sticking their hydrophobic tails out toward the air. Soap films and cell membranes share some deep similarities—worth keeping in mind the next time you wash your dishes.

TUBERCULOSIS AND TOUGH MEMBRANES

In London at the start of the nineteenth century, 30% of all deaths were due to tuberculosis, an infectious disease that most commonly affects the lungs. The "white plague" infected and killed vast numbers of people worldwide, and remained the first or second leading cause of death each year in the United States through the first decade and a half of the twentieth century. Even now, about one million people die of tuberculosis annually. Tuberculosis is caused by a bacterium: *Mycobacterium tuberculosis*. A different mycobacterium, *Mycobacterium leprae*, causes the disease leprosy, another scourge of humanity that has eaten away at the skin and nervous systems of its victims for millennia, until the advent of modern antibacterial treatments. The mycobacteria are notoriously tough. We've known for about a century, for example, that *Mycobacterium leprae* and *Mycobacterium tuberculosis* can survive periods of dehydration lasting several months. This is puzzling not only because cells need water for their biochemical activities but also from a membrane-based perspective: if there isn't water near the lipids' hydrophilic heads, how can hydrophilic and hydrophobic interactions hold a membrane together, and how can the membrane hold the bacterium together?

The mycobacteria, it turns out, have very odd membranes. Their interiors are bounded by a lipid bilayer, just as in your cells and the cells of every other organism. Outside this bilayer, however, is a dense hydrophobic gel, outside of which lies a lipid monolayer—in other words, a single layer of lipids, whose greasy tails point inward and whose hydrophilic heads face the outside. Not only is this arrangement of lipids unusual, but the lipid molecules are unusual in themselves: many have a sugar called trehalose bound to their hydrophilic heads (green in the illustration). The mycobacteria and a few closely related soil bacteria are the only organisms on the planet known to have trehalose lipids. Could this be important?

I learned about these mycobacterial membranes about a decade ago, not long after starting my research lab at the University of Oregon.

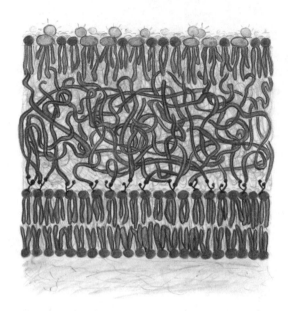

I had been working with lipid bilayers for a few years by then, mostly measuring physical properties, such as stiffness, to understand what these materials are capable of. Through some experiments involving sugars and polymers, I began to work with the group of Carolyn Bertozzi, a chemist then at the University of California at Berkeley. Coincidentally, one of her lab's major projects was to unravel the methods by which the mycobacteria create trehalose lipids and other strange molecules, both to understand the chemical tools nature has developed and to potentially disrupt them to tackle disease. It was through these interactions that I first heard about trehalose lipids, which immediately rang bells because trehalose, in other contexts, is an almost magical sugar. There are a handful of organisms, including some fungi, plants, yeast, and even certain animals, that can survive the loss of over 99% of their water. The "resurrection plant" *Selaginella lepidophylla*, for example, can withstand near-total dehydration for years, curling into a tight, brown ball that revives as an ordinary-looking green plant when hydrated. A common feature of many of these organisms is that they make trehalose, often in copious amounts. Compared to other sugars, like the familiar glucose and sucrose, trehalose is less likely to form crystals when concentrated, leaving the sugar molecules more available to interact with other substances. In addition, trehalose readily forms so-called hydrogen bonds, similar to the bonds between water molecules and between water and hydrophilic molecules, allowing the sugar to mimic water to some degree. Unlike water, however, trehalose doesn't readily evaporate. It's believed that for all these reasons, trehalose is a useful substance for dehydration resistance. In fact, there's a lot of contemporary effort to put it to use outside of organisms to preserve cells such as blood cells, and biomaterials such as proteins and vaccines, in a dry state for storage and transport. Perhaps the mycobacteria, I wondered, adapted the trehalose tool kit to link it to lipids in order to protect their outer membranes from dehydration. How could we test this idea?

We couldn't just engineer the microbes to stop making trehalose lipids and then test their resilience because we don't understand the mycobacterial machinery well enough to alter it. Even if we could, I wasn't keen on keeping tuberculosis-causing bacteria in the lab. (As we discuss later, my group happily works with the bacteria that cause cholera, but cholera is easy to prevent, hard to catch, and easy to cure— the opposite of tuberculosis.) I decided on a different approach instead, one that's been very fruitfully applied to "normal" lipid bilayers for decades: re-creating artificial, cell-free membranes on solid surfaces. Normal lipids can be coaxed to form bilayers on very clean and flat glass surfaces. Moreover, because of the hydrophilic nature of the glass and the lipids, a water layer just 1 to 2 nanometers thick separates the lipids and the glass, allowing the bilayer to retain its two-dimensional fluidity. At the cost of sacrificing some of the realism of an intact cellular membrane, this provides a controllable, convenient platform for studying bilayer biophysics.

We decided, therefore, to try to build an analogous supported membrane platform to mimic the non-bilayer structure adopted by tuberculosis lipids. We first chemically linked hydrophobic molecules to glass wafers. Then we formed monolayers of lipids at the surfaces of water-filled troughs, where the hydrophobic tails would stick out into the air, and gently transferred these monolayers onto the wafers so that the tails met the linked hydrophobic layer. We constructed the monolayers to contain specific fractions of purified trehalose lipids, among other more conventional lipids. Just like the natural mycobacteria have a dense, hydrophobic layer underlying a monolayer that contains trehalose lipids, our artificial membranes had a dense, hydrophobic layer underlying a monolayer that contains trehalose lipids.

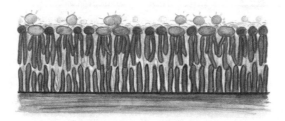

With this platform in hand, we could dehydrate and rehydrate the supported membrane. As expected, lipid monolayers composed completely of "normal" lipids didn't survive desiccation. In contrast, monolayers composed almost entirely of the mycobacterial trehalose lipids were intact after dehydration and rehydration, and even retained their two-dimensional fluidity. More remarkably, monolayers that were a mixture of normal and trehalose lipids could survive the loss of water down to a trehalose lipid concentration of about 25%. Even a minority of trehalose lipids can make a membrane resistant to dehydration! We pushed this a step further with our colleagues in the Bertozzi Lab, especially David Rabuka, who created synthetic trehalose lipids with the same hydrophobic chains as more standard lipids, differing only in the sugar at the head. (The natural mycobacterial lipids have gigantic hydrophobic chains. One could imagine these chains intertwining in some way, with this entanglement rather than trehalose driving the membrane preservation we observed.) These artificial molecules saved membranes from desiccation just as well as the mycobacterial lipid, confirming the notion that the trehalose itself is the protective agent. It was a satisfying result for our colleagues, me, and my nascent research group.

It seems that the bacteria that cause tuberculosis and leprosy have figured out a clever and robust route to stress resistance by linking sugars to lipids and of course exploiting the self-assembly of lipids into membranes to define their exterior. Could we engineer even better artificial dehydration-resistant lipids, for example, with multiple trehalose sugars per molecule, to construct easy-to-store biomaterials? Could we destroy lipid-linked trehalose to counter tuberculosis? I don't know; we'll see what the future may bring.

ORGANIZING A TWO-DIMENSIONAL LIQUID

Returning to normal cellular membranes, the two-dimensional fluidity of lipid bilayers presents a potential problem for cells: How can a cell organize its membrane, housing certain proteins together with their partners and keeping other proteins separated, if the membrane as a whole is a liquid? One tactic—taken, for example, by the T-cells described above—is to link membrane proteins to the internal scaffolding of the cell, whose struts and motors can push and pull as needed. Another possible tactic, one that's intrinsic to the physical properties of the membrane itself, involves two types of lipids. Both form fluid bilayers, driven by the goal of shielding their hydrophobic tails from water, but each prefers to be near its own type, A lipids next to A and B lipids next to B. Like oil and water, these two lipid types segregate, but the segregation is confined to the two-dimensional world of the bilayer. In the last decades of the twentieth century, scientists realized that this lipid membrane segregation could easily occur for mixtures of lipids commonly found in cellular membranes. The hydrophobic tails of different lipids can be relatively stiff or floppy, depending on the types of chemical bonds that connect their atoms. Combinations of stiff- and floppy-tailed lipids together with cholesterol (abundant in cell membranes) form bilayers that are a hodgepodge of two distinct compositions, each coexisting with the other. One composition is rich in cholesterol and stiff-tailed lipids; the other is rich in floppy-tailed lipids. Their segregation shows all the hallmarks of *phase separation*, which physicists had been studying for decades, especially its dependence on temperature. Above some critical temperature, the different lipids mix together (upper illustration, following page), while below that temperature they follow their preferences and segregate (lower illustration, following page).

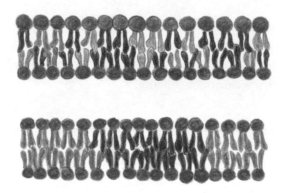

As is the case for DNA melting (chapter 1), the transition is a sharp one, and the analytic tool kit originally developed for nonbiological materials again finds applications in the living world. The picture that has emerged is that cells could make use of this cholesterol-dependent phase separation to organize their membranes. Proteins that prefer cholesterol-rich phases would be sorted into regions with other proteins of similar preference; cholesterol-poor regions would house other sets of proteins. In artificial lipid membranes, it's easy to see and study lipid phase separation. One can readily construct in the lab cell-sized spheres of lipid bilayer, which serve as tools for studying the biophysics of membranes and membrane proteins. (This is reminiscent of a soap bubble, but with water instead of air on both the inside and outside, and a single lipid bilayer as the boundary between them.) Labeling the membrane with different dyes that prefer cholesterol-rich or cholesterol-poor domains, for example, light and dark gray in the illustration, we see through our microscopes disks of one color amid a sea of the other.

It quickly became clear that this spatial organization *could* happen in actual cells, but whether it actually *does* turned out to be a difficult and contentious question that is still not fully resolved. In artificial membranes, cholesterol-rich and cholesterol-poor domains grow to sizes that are easy to see in a microscope. Moreover, one can lower and raise the temperature and watch domains appear and disappear at will. In living cells, puzzlingly, one never sees these domains, though we know that the lipids and cholesterol that make them up are the same

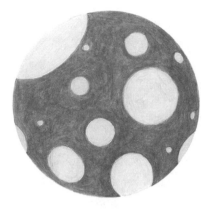

sorts used in the artificial membranes. The suggestion has been that the domains exist but are constrained by the cell's underlying scaffolding to be at most a few tens of nanometers in size. Due to the wavelike nature of light, we can't see structures smaller than a few hundred nanometers, rendering such domains unobservable. This is obviously not very satisfying; to say that something exists but is unobservable doesn't bolster one's confidence that it actually exists! Intriguingly, however, one can chemically perturb cells to create "blebs"—blisters of outer membrane detached from the underlying scaffolding. In these, one finds visibly discernible lipid phases with all the hallmarks of liquid phase separation, an observation first reported by Watt Webb and colleagues at Cornell University in 2007.

The bleb experiments lend credence to the idea that real membranes do phase-separate, but one might still argue that the cell's membranes are perhaps harshly perturbed and are not natural. Recently, scientists have observed large, visible domains in the membranes that bound an organelle called the *vacuole* in yeast cells. A group led by Sarah Keller at the University of Washington showed that these membranes in living yeast cells exhibit the hallmarks of phase separation, most importantly a critical temperature below which the domains appear. Intriguingly, the yeast cells seem to use these domains to enable the digestion of stored fats when their preferred sugars are unavailable, using the domains to concentrate the proteins involved in this process. It remains to be seen whether other cells have similar strategies, but the idea that cells harness the power of liquid phase separation to organize their membranes increasingly seems not only elegant but also true.

MEMBRANE STRUCTURE AND SELF-ASSEMBLY

This picture of cell membranes as two-dimensional fluids made possible by the self-organized lipid bilayer was cemented in the early 1970s, following decades of study of the nature of cellular membranes. The elegance of the lipid bilayer architecture is amazing: not only does it explain much of the behavior of membranes, but it shows that this behavior is a consequence of simple physical interactions. One might have expected that the tremendous biological importance of membranes would mean that cells would carefully specify the arrangements of lipids, creating precise chemical bonds to bind them together. But this isn't so: lipids are free to do as they please, just as a drop of oil suspended in water is free to form itself into a cube or a jagged star. The drop forms a sphere, not a cube or a star, simply because a sphere is the shape that minimizes the contact area between the oil and the water. Similarly, the lipids arrange themselves into a bilayer simply because that shape minimizes contact between the hydrophobic chains and water. The cell doesn't need genes to tell lipids to form bilayers; this is simply what lipids do. (The cell does need genes to encode the proteins that synthesize lipid molecules; once made, however, the lipids can organize themselves.)

As with the folding of proteins, we see the powerful general principle of self-assembly at work: simple physical concerns can orchestrate the formation of structure, allowing molecules to organize themselves. Harnessing self-assembly is not only useful for nature but also inspirational for those of us who study nature, showing that life need not be as complicated as it may initially seem—that physical simplicity may underlie biological complexity.

We've now met the essential molecules that make up every organism on earth: DNA, RNA, proteins, and lipids. This isn't a complete set of the ingredients of life—there are ions, sugars, hormones, and more that contribute in important ways—but understanding this group of universal components tells us a lot about how life works and how the living world encodes information. These molecules assemble and interact in exceptionally varied ways to generate the diversity of life around us. We'll continue to explore different types of biological structures and the physical forces that guide and constrain them, but we first dive into a major biophysical theme we've so far only hinted at: predictable randomness.

Predictable Randomness

Nothing is ever still. Every picture we've seen of a protein, DNA, or any other molecule is fundamentally flawed. Every lipid, for example, should be depicted as a blur of motion rather than sedately posed.

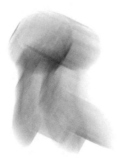

This motion is not unique to biological molecules. If I look into my microscope at a tiny glass bead about the size of a bacterium, adrift in a dish of water, within a few seconds it will meander over a distance of a few times its diameter. This travel isn't caused by flows in the water, or unevenness of my microscope stage. It's an intrinsic, unavoidable dervish that all objects perform. We see in this chapter that this motion—a consequence of fundamental physical laws—creates a backdrop to all of life's processes that is deeply alien to our macroscopic intuition, a backdrop that is fundamentally governed by randomness. Paradoxically, there is structure in this chaos, and many of the machineries of life make sense once we grasp how randomness and predictability are interrelated in the small-scale world.

THE PHYSICS OF POLLEN

This ceaseless dance of small things is called *Brownian motion*, after the botanist Robert Brown who observed and described it in 1827. Brown looked though his microscope at granules inside grains of wild-flower pollen and saw their incessant motion. The motion is random: on average, there are as many steps to the right as to the left, upward as downward, with no pattern to the sequence of directions. Other mi-croscopists had seen these dynamics before, but Brown established that they were not a consequence of life, the biological origin of their constituents, currents in the fluid surrounding the particles, or flows induced by the evaporation of liquid; rather, they were due to some sort of universal, underlying physics. To probe, for example, whether evap-oration was a driver of the motion, Brown shook up a mixture of oil and pollen-containing water, with water droplets thereby shielded from evaporation by the surrounding oil; the granules nonetheless showed clear Brownian motion.

Brown established that everything incessantly shakes and wanders. But *why*? What drives this microscopic motion? Many decades passed without an answer, until Albert Einstein in Switzerland, Marian Smo-luchowski in Poland, and William Sutherland in Australia each inde-pendently came up with a clear, simple, and accurate explanation in the first few years of the twentieth century. Their first insight was to take seriously the circumstantial evidence collected throughout the nineteenth century, especially by chemists, that matter is made up of discrete units (called atoms). This seems trivial from a modern point of view—we're so used to referring to atoms and molecules that it seems strange not to—but at the turn of the twentieth century, the notion of discrete building blocks for matter, rather than an endlessly divisible continuum, was contentious and not universally accepted. Einstein and others pointed out that the jostling of lots of tiny, individual water mol-ecules colliding with Brown's granules or my glass spheres would give rise to motions with exactly the same form observed in experiments.

The second insight was in realizing the role of temperature. We're permeated by something called *thermal energy*, of which temperature, roughly speaking, is a measure. The greater the temperature, the more thermal energy an object has; the lower the temperature, the less thermal energy. As Einstein, Smoluchowski, and Sutherland realized, combining the impetus of thermal energy with the viscous drag provided by the fluid surrounding an object leads to a predictive model of random motion that matches perfectly with experimental observations. What's more, the underlying principles are universal and unavoidable; where there is temperature, there is random motion. (Only the unattainable temperature of absolute zero, −273.15°C or −459.67°F, brings stillness.) To appreciate the biophysical implications of this model, we need to be a bit more precise in describing Brownian motion. We've said that it's random. Nonetheless, it is comprehensible.

QUANTIFYING RANDOMNESS

Suppose that you walk in a straight line for 10 seconds and cover a distance of 30 feet. If instead you walk for 20 seconds, you wouldn't be surprised to have traveled 60 feet; for 100 seconds, 300 feet, and so on. We say that distance is proportional to time; to double your distance, you'd need to double the time spent walking. A graph of distance plotted against time would look like a straight line. The slope of the line would be a measure of your speed (in this case, 3 feet per second).

Sketching the paths my meandering microscopic bead might take over 10 seconds shows a variety of tangled tracks. We can't predict its

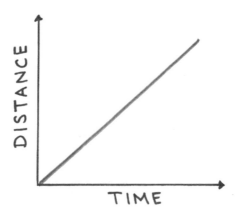

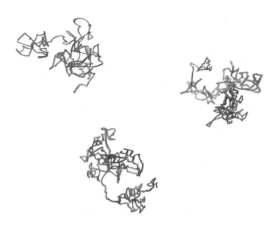

route ahead of time, nor what its final location will be. Its motion is random.

Coexisting with this randomness, however, is a sort of predictability. Just as I can't tell in advance which face a flipped coin will land on, but I know that if I flip the coin many times roughly half will show heads and half tails, I can make statements about the statistics of Brownian motion. If I repeat the 10-second meandering of my bead a few dozen times and draw a point at each of the final locations, with the starting position at the center of the page, the set of endpoints would look like the pink cloud I've illustrated.

Though the positions are random, the average distance from the starting point is well defined. How does this average depend on the travel time?

If you're experiencing déjà vu, that's great! This is, in fact, equivalent to the problem we encountered when asking how large a blob of DNA is in chapter 3. There we met the random walk, and learned that a random walk of N steps ends up, on average, the square root of N step-lengths away from its starting point. Here every instant of time during which our Brownian particle is bombarded by atoms of the fluid gives the particle a random kick, from which it takes a random step. Therefore, on average, the distance the particle covers is proportional to the square root of the time it travels. A graph of the typical distance versus time, rather than being a straight line, is curved.

If the particle travels four times longer, it will, on average, only go twice as far. To move three times as far (on average), the travel time must be nine times longer.

In addition to time, Brownian motion also depends on the size of the particle. This makes sense—after all, we've stated that random motion is significant for microscopic particles, and we know that we don't see larger objects, like watermelons and baseballs, randomly shaking across the floor. All particles have an average displacement that grows with the square root of time, but the magnitude of this growth is greater for smaller particles than larger ones. All particles get the same kick from the ambient thermal energy, but the smaller particles respond more vigorously to it.

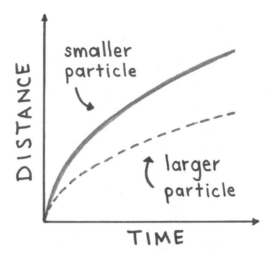

The random motion of molecules is also referred to as *diffusion*, a term commonly encountered in descriptions of dyes drifting through a liquid or gases wafting through the air. I note, however, that the common classroom demonstration of perfume spreading through the air isn't really an illustration of diffusion. The perfume molecules are certainly undergoing Brownian motion, but the main drivers of their motion across a room are air currents set up by nonuniform temperatures or by ventilation systems, people moving around, and other disturbances.

HOW TO BUILD SMALL THINGS

Beyond the obvious consequence that salts and sugars, lipids, proteins, and even whole cells are in constant agitation, Brownian motion illuminates a great many aspects of biology. First of all, it solves a nagging problem with our discussions of self-assembly in earlier chapters. As we've seen, proteins fold themselves into particular three-dimensional shapes, driven by physical interactions between the amino acids that make them up. Lego bricks, however, also have particular interactions between them, but a pile of bricks doesn't spontaneously assemble itself into some form. Brownian motion explains the difference. Being small, the amino acid chain is in constant, vigorous motion. The molecule is always jiggling about, placing some amino acids in close proximity to others, then others, then others still, until it settles into a configuration in which sufficiently strong interactions lock it into place. Similarly, thermal energy drives the random motion of lipids; they find each other and assemble into a membrane. The recipe for self-assembly, therefore, is not merely physical interactions, but physical interactions together with Brownian motion.

Similarly, gene expression and regulation also depend on Brownian motion. We've described transcription factor proteins binding to DNA, glossing over the question of how the proteins find their target DNA sequences. There's no guiding hand or train track conveying them smoothly to their destination. Rather, buffeted by thermal energy, the proteins wander through the cellular space, colliding with all sorts of DNA regions and being held for a while by those they specifically match. As with self-assembly, this strategy for running a machine wouldn't work for a macroscopic object—I can't set my office key on the floor and hope it wanders off and finds the door lock—but it's a great strategy in the microscopic world.

WHAT SETS THE SPEED OF THOUGHT?

Brownian motion also illuminates a deep connection between structure and time. As a concrete example, let's think about the connection between two neurons.

There are two types of connections. In one, called a *chemical synapse*, the two cells are separated by a space a few tens of nanometers across. (Recall that a nanometer is a billionth of a meter.) The cells communicate with each other by sending chemicals called *neurotransmitters,* the small orange circles in the illustration, across this cleft.

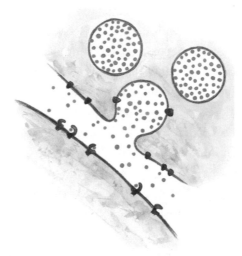

There are many different neurotransmitters and many drugs and pharmaceuticals that target their release, uptake, or degradation. Nicotine, for example, and some of the drugs used to treat Alzheimer's disease increase the levels of the neurotransmitter acetylcholine. Adenosine, another neurotransmitter, slows down brain activity, making you drowsy; caffeine blocks adenosine's receptor protein from binding it, keeping you awake. How do the neurons manage to send and receive neurotransmitters across a chemical synapse? By doing nothing more than releasing these chemicals and then letting diffusion do the work of spreading them. Left alone, the molecules meander about the cleft, occasionally running into receptor proteins on the target cell that bind them and trigger a neural response. There's no machinery needed—no nanoscale delivery van, no electromagnetic forces pushing the neurotransmitters along. The neurotransmitters are small, a nanometer or so in size, and their vigorous Brownian motion carries them a few tens of nanometers within about a microsecond.

Turning this explanation around, we can ask how long it takes information to be transmitted across a chemical synapse. If one neuron is stimulated so that an electrical signal travels along it until reaching the end of the cell, the news of its stimulation must be conveyed to the next cell in the relay—perhaps another neuron, as above, or perhaps a muscle cell. This intercellular baton passing takes about a microsecond, as we've seen—a millionth of a second. This is a rough estimate, of course. Strictly speaking, we should ask about how long it takes some threshold number of random walkers to cross the cleft rather than the average molecule. Still, whatever the criterion, the timescale is microseconds. There's no reason to expect it to take much longer—thousandths of a second, for example—and no physical way it could be much faster—billionths of a second, for example—given the physical size of the synapse.

Since I was a child, I've wondered what sets the speed of thought—why a minute feels like a minute and not a year, why we can't draw out and savor each millisecond of our experience. At a chemical synapse, Brownian motion determines the communication speed between neurons in an inescapable way. There are several other ways information can be transmitted in the brain, each with its own dynamics. Every biological information pathway, however, is governed in some way or another by molecular flows, of which Brownian motion is an integral part, helping set the rate at which our brains function.

The microsecond timescale of a chemical synapse is quite fast—obviously adequate for our needs. It's interesting to contrast this, however, with the timescale of modern computers, which is around a nanosecond, a billionth of a second, per operation. My laptop operates far faster than my brain. Rather than relying on molecules, it harnesses the motion of far smaller electrons and, moreover, directly pushes them around with electric fields. My brain is relatively slow, but its neurons are much more interconnected than the transistors in my laptop's central processing unit. The neural architecture enables a dazzling number of calculations to be performed in parallel among different sets of neurons at the same time rather than in a rigid, temporal sequence. Both the connectivity and the parallelism are helpful for conceptually difficult tasks. It's interesting to imagine what will happen when machines surpass us in both computational speed and network complexity; we'll likely reach that point soon.

CARGO TRANSPORT IN CELLS

In the example above, a neuron simply lets go of neurotransmitters, assured that they'll diffuse to their target in a reasonable time. Cells of all sorts rely on Brownian motion in similar ways. Recall from chapter 4 our bacterium that loves lactose: the lac repressor protein may or may not encounter lactose that the bacterium has picked up from its environment, and this controls whether it will bind to a particular DNA segment and inhibit creation of lactose-digesting proteins. How does the lac repressor find the DNA? Again, nothing in particular directs it. The protein meanders at random. Given its size, its random motion is vigorous enough that it will wander about 1 micrometer, the width of a typical bacterium, within about a hundredth of a second. Reaching a particular destination—for example, the repressor's specific target DNA sequence—takes longer, since most random paths don't go anywhere useful. Still, it takes only about a tenth of a second on average to reach any given point. Therefore, it wouldn't surprise us that a bacterium can reasonably make decisions informed by its environment within a few tenths of a second.

Imagine, in contrast, a typical eukaryotic cell—one of your body's white blood cells, for example, which is about 10 micrometers wide, 10 times wider than a typical bacterium. By Brownian motion, a protein would need 10 squared, or 100, times as long to diffuse a distance equal to this cell's width. Finding a specific target like a DNA binding site is even more challenging. The average time required, it turns out, is roughly proportional to the cell size cubed, and so would take $10 \times 10 \times 10 = 1000$ times as long in the white blood cell as in the bacterium. Rather than a tenth of a second, the timescale would be nearly 2 minutes—a sluggish response!

Rather than acquiesce to lethargy, eukaryotic cells take a more active approach, using motor proteins to ferry cargo. We've seen one of

these motor proteins, kinesin, already; it grabs lipid-and-protein encap-
sulated material with one end and marches along microtubule roads
with the other.

A typical kinesin walks with a speed of about 2 micrometers per second,
so it could travel the length of a eukaryotic cell in a few seconds—much
faster than the minutes required for diffusion. Even here, though, the
cell still exploits randomness: the motor protein need not ferry its cargo
all the way to its destination, but simply close enough that Brownian
motion can rapidly take care of the last stage. (Upon reaching the nucleus, for example, from a starting point far away in a large cell, random
diffusion can quickly bring a transcription factor to its DNA target,
about a micrometer away.) The utility of molecules like kinesin is clear,
but there is also a cost: the cell must expend energy to power motor
proteins rather than relying on the Brownian motion that the universe
provides for free.

So far, despite lots of searching, no one has found kinesin-like motor
proteins in prokaryotic cells (bacteria and archaea). From a biophysical perspective, this makes sense: it's not that bacteria couldn't have
evolved them, but rather that they don't need to. Brownian motion is
fast for small things and slow for large things. Nearly all bacteria are
small, and so they can simply and efficiently rely on this randomness
for their internal transportation needs.

WHY DO BACTERIA SWIM?

Transportation outside of bacteria, including the motion of the bacteria themselves, is also deeply connected to randomness. Most bacteria can move, for example, by swimming through liquid. *Escherichia coli*, for example, has several whiplike flagella that spin one way to propel the organism forward and spin the other way to make it turn. These microbes are constantly in motion, and with a microscope one can watch them darting and dashing in a dish of water.

We might think that the bacteria swim to gobble up food, like miniature baleen whales scooping up krill in their path, but physics reveals that this is not the case. *E. coli* swims with a speed of about 10 micrometers per second, so if there were some food about a micrometer away (a distance similar to their body length), it would take a tenth of a second to swim to it. Their "foods" are sugars and other small molecules, less than a thousandth of a micrometer in size, so small that they would take only a millisecond or so to meander 1 micron. If you're a bacterium, food diffuses to you much faster than it would take you to swim to it! As physicist Edward Purcell put it, "You can thrash around a lot, but the fellow who sits there quietly waiting for stuff to diffuse" does just as well.

So why swim? Bacteria like *E. coli* measure the concentration of food in their surroundings, counting the rate at which nutrient molecules hit their receptors, and they swim toward regions of higher concentrations. Purcell again: "What it [the bacterium] can do is find places where the food is better or more abundant. That is, it does not move like a cow that is grazing a pasture—it moves to find *greener pastures*." Thanks to many years of work, we now understand *E. coli's* sensing and decision making in exquisite detail: how the detection of nutrients subtly alters the flagellar control proteins so that the little creatures swim longer on straight paths if they're in regions of increasing nutrient richness, and tumble more to change their orientation if they're headed in a less promising direction. The same sorts of mechanisms are at work in all sorts of bacteria, including many that navigate into animals, and also in many eukaryotic cells, for example, immune cells that migrate toward wound sites.

We've now met many of the pieces that make up cells, and the physical processes and motifs that orchestrate their assembly, their decision making, and their dynamics. Cells are, of course, wonderful—living, growing, reproducing entities that exist in bewildering numbers, trillions in each of us alone. Cells can be even more amazing if working together. In part II, we expand our scope to ensembles of cells, including embryos, organs, bacterial communities, and whole organisms of all shapes and sizes—again seeing general biophysical themes at play as interacting cells self-assemble, make decisions using biological circuitry, deal with randomness, and scale themselves to larger sizes.

PART II

Living Large

Assembling Embryos

We've now met the major building blocks of life, and we've gotten to know three of the overarching themes that govern their interactions: the concept of self-assembly, the predictable randomness of microscopic motion, and the construction of regulatory circuits. We've caught a glimpse of the fourth theme, scaling, reflected in the size dependence of Brownian motion and the long time needed to diffuse large distances. We focus much more on scaling a few chapters from now.

So far, most of our examples of these themes have been at the level of single cells or their internal machineries. These ideas also apply, however, to beating hearts, bananas, three-toed sloths, and all other manifestations of life at larger scales. Biophysical motifs shed light on collections and communities of cells, including whole organisms, and we'll uncover among them elegant examples of simplicity underlying complexity.

"THE TOTALITY OF ALL PRIMORDIA"

Rather than thinking first about groups of a few cells, or particular tissues or organs, let's fearlessly dive into what is arguably the most complex and amazing phenomenon of the living world: the development of an animal, like you, from a single fertilized egg cell. Our understanding of embryonic development has itself developed quite dramatically. Just a few centuries ago, it was widely believed that this single cell held within it a *homunculus,* a miniature but fully formed human from which the baby, child, and adult grew. In fact, some early microscopists convinced themselves that they saw these little people through their instruments' eyepieces, preformed in sperm or unfertilized egg cells. We

know now that the single-celled embryo simply contains a genome—
DNA from the mother and father—along with proteins, RNA, and other
useful ingredients donated mostly by the mother. From this starting
point, the cell divides, and divides, and divides. The descendant cells
don't just split and grow, but change their positions, shapes, sizes, and
gene expression patterns until reaching the set of positions, shapes,
sizes, and gene expression patterns of a functioning organism.

Even with science as our starting point, the transformation from cell
to animal can seem magical. Let's step back just over a century, to the
late 1800s, when many of the pioneering experiments of experimental
embryology took place. By watching animals develop and also by prod-
ding, splitting, and transplanting, scientists began to map out the
paths by which cells obtained particular identities and tissues acquired
form. One of these pioneers was Hans Driesch, a German biologist
working mostly in Naples. Driesch discovered that separating the cells
of a two-cell sea urchin embryo resulted in each of the individual cells
growing into a normal sea urchin. Even separating cells from a four-
or eight-cell embryo would also often lead to a single cell growing into
a full animal. What's more, Driesch found that gently pressing a young
embryo could move cells from their standard positions (with cells that
would normally be the progenitors of the top part of the animal being
instead at the bottom), and they remained there when the pressure was
released. Despite the shuffling, the sea urchin developed normally, as if
the transported cells knew their new locations and acted accordingly.
Each cell, Driesch concluded, "carries the totality of all primordia," a
perspective very much at odds with a simple mechanical view. One
can't scramble the gears of a watch or the pistons of a steam engine
and find that they possess some deep, inherent knowledge of the new
roles they need to take to make a functional machine. Driesch was so
struck by the apparent contradiction between the workings of an embryo and the physics at his disposal that he abandoned the study of
development altogether to become a philosophy professor, promoting
the view that living things are governed by laws fundamentally different than those of nonliving substances.

Even then, Driesch's leap was extreme. In contrast, biologists such as the American Ross Granville Harrison advanced the notion that factors intrinsic to each cell and factors more broadly dispersed through the embryo together orchestrated development, a view consistent with our modern perspective and since fleshed out by a hundred years of work.

Before you get your hopes for this chapter set too high, anticipating that we will unveil the full path from single cell to complex creature, I note that embryology is not a solved problem. We can't take your genome and predict just from the sequence of As, Cs, Gs, and Ts that you're a two-armed, two-legged, hairy, air-breathing animal. We can't predict, given just the genome of a starfish, that the animal will progress from a soft, free-swimming two-fold-symmetric larva to a hard, typically five-fold-symmetric predator stalking seabeds and rocky shores. In fact, if we didn't know the organism from which the genomic DNA was taken, we could only tell that the starfish genome codes for a marine invertebrate and the human genome codes for a primate by comparison to other known genomes, not by modeling from first principles the activities of all the proteins and regulatory networks encoded by the constituent genes. Nonetheless, we can say a lot about development, thanks especially to two key features.

First, genes are remarkably similar across organisms, so learning what a particular gene does in an easily studied creature, like a mouse or a fruit fly, tells us a lot about what that gene does in another creature, such as a human.

Let's consider as an example the gene "sonic hedgehog," which encodes a protein crucial for the development of limbs and also active within expanding cancerous tumors. In a classic study published in 1980, Christiane Nüsslein-Volhard and Eric Wieschaus discovered several genes that are important determinants of the body plan of the fruit fly, naming one "hedgehog" because mutations in it give rise to spiky fly larvae. Similar genes were later found throughout the animal kingdom. Mammals, including humans, have in their genomes three genes that are very similar to the fruit fly's hedgehog. Two, desert hedgehog and

Indian hedgehog, are whimsically named after species of actual hedge-hogs. The third is sonic hedgehog, even more whimsically named after a speedy blue video game character; one of the researchers involved was inspired by a *Sonic the Hedgehog* comic book belonging to his daughter.

All these proteins are remarkably similar. I've depicted the structure of part of the fruit fly's hedgehog protein (left) and part of the human sonic hedgehog protein (right). Each has a pair of tilted helices, a few short sheets, and various loops connecting them, all arranged nearly identically.

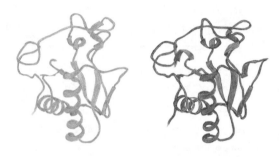

It's easy to tell a fruit fly from a person, but it's very hard to tell their hedgehog proteins apart. This similarity is also evident if we list the amino acids that make up each of these chains. I'll just write here a stretch of 46 amino acids, about a third of the chain I've illustrated, using a common alphabetical code in which each letter denotes a different amino acid, with shared units in bold:

Fruit fly:

RCKE**KLN**V**LA**YS**VMN**EW**PG**I**RLL**V**TE**SW**DED**YHHGQ**ESLHYEGRAV**

Human:

RCKD**KLN**A**LA**IS**VMN**QW**PG**V**KLR**V**TE**GW**DED**GHHSE**ESLHYEGRAV**

For sequence as for structure, the resemblance is striking. For the complete protein, about 70% of amino acids are identical between the fruit fly hedgehog protein and the human sonic hedgehog protein. Even the mismatches aren't as different as one might think. In the list above, the first discrepancy is an E (glutamic acid) in the fruit fly protein and a D (aspartic acid) in the human protein; both are negatively charged. The V and A in the next mismatched pair are both hydrophobic (valine and alanine, respectively). Even if the exact molecular identities of the amino acids differ, their physical attributes are in many cases similar. Nature's parsimony allows us to amplify the impacts of learning about its tools; we can be reasonably confident that the behavior of hedgehog in fruit flies is similar to the behavior of sonic hedgehog in humans, or desert hedgehog in desert hedgehogs.

The second reason we can gain general insights into development is even deeper: nature makes use of robust physical mechanisms to pattern and organize cells. These mechanisms, just like the genes and proteins involved, are also common across organisms. Let's see how they work.

KNOWING WHERE YOU ARE

Different organs grow at different locations. Wings emerge near the middle of a mosquito and antennae at the head. Your fingers sprout at the far end of your hand, not at your wrist. One might imagine that only dedicated wing-forming cells migrate toward and end up at a wing-forming zone in the middle of a developing insect—in other words, that the cells' fate is specified prior to their positioning. Or one might imagine that cells throughout the body are capable of forming wings, but only those at the appropriate location get a signal instructing them to do so. Nature, it turns out, uses both of these tactics. The second, in which spatial cues guide cell fate, is surprisingly common, and it enables an efficient encoding of instructions for the developing organism.

The existence of spatial cues has been known for over a century. Experiments like Driesch's in which the arrangements of cells were deliberately scrambled in the embryos of sea urchins and other animals, or in which some cells were transplanted from one region of one animal into a different region of another, nonetheless often lead to normal development, as if the scrambled or transported cells knew their new addresses in the embryonic neighborhood and acted appropriately. Understanding this seemingly magical sensory ability and revealing the nature and consequences of spatial cues is a more recent and still ongoing story. The basic idea, however, is straightforward and involves two of the biophysical machineries we've already met: random diffusion and regulatory networks.

The sonic hedgehog protein we discussed above isn't distributed uniformly throughout an embryo, nor is it present at some fixed concentration in some regions and absent in others. Rather, it exhibits a concentration gradient, progressing smoothly from high values around where the protein is being made to progressively lower values farther away. (Like all proteins, it decays, so the total amount of protein isn't constantly increasing.) This gradient is simply the consequence of diffusion, the random walk of molecules from their starting points, smearing out a cloud of molecules as we saw in chapter 6. Sonic hedgehog is produced at many sites in developing organisms, giving rise to many local concentration gradients. One location is the limb bud, the early precursor to each limb, which takes shape in week three of human embryogenesis. The sonic hedgehog distribution at the limb bud is most concentrated at one side, diminishing toward the other.

If you look at your left hand, palm facing you and fingers up, your thumb is on the left and your little finger on the right. Though we've never met, I can say with high confidence that this is the case, and that the ordering of your fingers isn't reversed or random. This arrangement is a consequence of the sonic hedgehog gradient: where the concentration of this protein is high, the little finger forms; where it is low, the thumb. The same holds in other animals. In the wing bud of a chick, the sonic hedgehog gradient governs the ordering of three bony digits, which progress 3-2-1 along the diffusing concentration profile. (Top: The orange blur is the sonic hedgehog gradient; the bones are as they appear in a four-day-old embryo.) Transplanting tissue from the protein-producing region of one chick wing bud to the low-concentration side of another wing bud gives two mirror-image concentration profiles from which six digits grow, 3-2-1-1-2–3 (bottom). The cells simply read the local sonic hedgehog concentration, unaware of the strange manipulations that generated it. The chick wing experiment, performed by Cheryll Tickle and colleagues at the University of Bath, used pat-

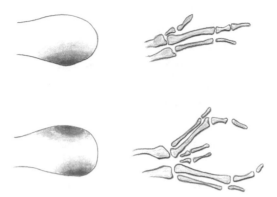

terns of digit development to probe which dinosaurs birds descended from, in addition to mapping the processes by which digits are derived. Embryonic cells measuring hedgehog protein concentrations is an ancient practice. Hedgehog gradients curate the array of fingers in your hand and the array of suckers on a cuttlefish arm, for example, though the last common ancestor of humans and cuttlefish lived over half a billion years ago. (Cuttlefish are not fish but rather cephalopods, closely related to squid and octopuses.)

Sonic hedgehog gradients drive the patterning of tissues other than limbs, playing a role in the formation of the nervous system, facial features, lungs, teeth, and more. The protein also turns up in cancers; cancer development often involves the activation of genetic processes associated with embryonic development, fostering rapid growth in unwanted ways.

Sonic hedgehog is just one of many *morphogens*, substances that govern the development of shape by variation in their concentration. Morphogens were predicted and named by mathematician and computer science pioneer Alan Turing in 1952, decades before any specific examples were discovered, in a prescient paper exploring the theoretical possibilities of such systems. Every developing embryo is crisscrossed by many coexisting and interacting morphogen gradients.

What are these morphogens doing? Either directly or via intermediaries, they act as transcription factors, turning on or off various genes, as we explored in chapter 4. The efficacy of a transcription factor depends on its concentration. This too is a consequence of physics: the binding of any molecule to any other is a constant flurry of attachment and detachment, and the probability that some transcription factor is bound to a DNA target is greater if there are more copies of the transcription factor floating around. The response function—the likelihood that the gene will be expressed, or the rate at which the protein encoded by the gene will be produced—can be a smooth function of the activator or repressor transcription factor concentration, or it can be sharp and switch-like, for example, nearly zero for low levels of an activator and at a high "on" state for activator concentrations above some threshold.

The concentration dependence of gene expression can allow surprisingly intricate patterning. Let's consider a stylized example now and add realism later. Imagine an elongated, pill-shaped embryo. This part isn't unrealistic—nearly every organism starts as a ball or an ellipsoid; we're all blobs at the beginning. Suppose the source of morphogens is at the left end, from some particular cells or materials supplied by the mother. Morphogen A diffuses, setting up a concentration gradient that decays along the length of the embryo (upper left illustration). If there's a switch-like response to morphogen A—on if A is abundant and off when it's rare—we have a steplike pattern of the output of that gene (upper right).

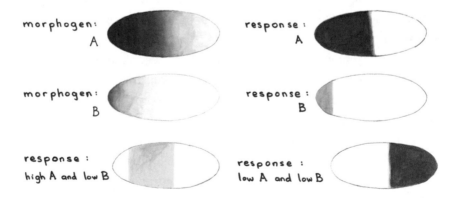

Now suppose there's also morphogen B, originating from the same location. Perhaps molecule B is larger and so its random walks aren't so vigorous; its concentration therefore has a steeper profile, and a similarly switch-like response would be localized to a smaller region.

Recall from chapter 4 that cells can create circuits of genes, integrating inputs to regulate their response. A circuit that is on if levels of A are high and B are low, and is off for any other combination, will therefore have an output profile that is on in a band just left of the middle. A "low A and low B" circuit would give a response in the right half of the embryo.

Even with only two transcription factor gradients, more than two spatial patterns for subsequent gene expression can emerge. An activated gene might encode a protein responsible for some developmental activity that should occur only in a particular region, or might encode another transcription factor that can itself interact with the first two. Suppose transcription factor C is regulated by the "high A and low B" circuit, and so is expressed in a wide band in the middle of the embryo from which it spreads by diffusion. A "high C and low A" circuit would then be activated in a narrow band, the zone just beyond where C is produced, but at which the diffusing C is still concentrated enough to be read as "high."

The narrow band demonstrates organizational precision, made possible by just a few genes. With still more transcription factors, each with an associated concentration gradient, the number of possibilities explodes. It's easy to imagine setting up specific patterns of gene expression precisely suited to the fates of the cells that will make up specific organs and tissues.

This may sound compelling in principle; remarkably, organisms actually adopt this approach in practice. We've known about such patterning in an approximate sense for decades, observing the blur of transcription factor concentrations and gene expression profiles, and in a much more precise sense in recent years for a small but growing number of genes and organisms. Our most exquisite understanding of embryonic spatial patterning comes from studies in fruit flies, again

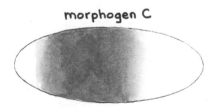

morphogen C

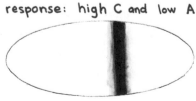

response: high C and low A

originating in the work of Christiane Nüsslein-Volhard and Eric Wieschaus. The early fly embryo, well before it develops legs, wings, or even
a head, is an elongated oval, just like our hypothetical embryo above.
The mother fly bequeaths RNA to her egg at one end, setting up an initial gradient of a transcription factor called bicoid that looks like our B
morphogen. Bicoid binds to the promoter region of a gene called hunchback and acts as an activator. The hunchback protein also shows a high-
in-front, low-in-back concentration gradient. About half a dozen other
genes follow, expressed in wide swaths, combinations of which enable
finer-scale expression patterns, like the seven stripes of the "even-
skipped" gene, emerging just three hours after the egg is fertilized.

The patterns of even-skipped and other genes set up the organization
of the fly body into fourteen segments. Different structures are fash-
ioned from different segments. Three segments, for example, form the
thorax, and each of the three grow a pair of legs, making use of
hedgehog and other genes.

It's especially satisfying that this specific pattern of stripes and segments, crucial to establishing the body plan of every fly ever formed,
can be both measured and predicted based on our biophysical understanding. Regarding measurement, one can survey in living embryos
the transcription of a particular gene into RNA by engineering into the
genome sequences that encode fusions of RNA-binding proteins and a
fluorescent protein like GFP (chapter 2). Expression of the gene, therefore, generates visible, glowing proteins. One can also monitor protein
levels themselves through similar insertions of fluorescent protein
genes. In either case, the fluorescent glow provides a precise, quantifiable reporter of the what, where, and when of developmental activity.
Regarding predictions, one can write down the equations of Brownian

motion and genetic response functions and calculate patterns of protein abundance and gene activity. These computational outputs aren't yet perfect mirrors of Nature's patterns, but they do reveal stripes and gaps with sizes and timescales that are admirable representations of the reality of the fly embryo.

Looking further, the fly shows us even deeper aspects of this embryonic patterning. In our hypothetical example, we considered just two levels of gene readout, driven by low and high morphogen levels. Cells could have more finely tuned senses, giving three different responses to three morphogen levels (low, medium, high), or four, or five, or more. Having more levels seems appealing: the organism could construct fine-grained structure from even fewer ingredients. What limits an embryo's precision? Could the fruit fly detect 1000 different bicoid concentrations at 1000 different locations along its body, setting up 1000 different anatomical features from just a single spatial gradient?

This too is a question for biophysics. The limits on the precision of patterning are set by the inherent randomness of diffusive motion (chapter 6) and the randomness of molecular binding, which we haven't explicitly examined but which has underpinnings similar to Brownian motion. One can easily say on average where a cloud of diffusing molecules will be, but how well this average represents the whole cloud depends on how many molecules are present. If I flip a million coins, I'm quite confident that the fraction that land heads will be very close to one-half. If I flip six coins, I wouldn't be surprised to find four heads and two tails. The embryo, similarly, could make lots of morphogen molecules and generate a smooth, well-defined gradient from which it could reliably discern many different concentration levels. Or it could make just a few molecules, requiring less effort and energy but resulting in noisier gradients that it could only coarsely read as regions of high or low concentration. Experiments suggest that the preferred strategy is closer to the latter than the former; there aren't millions of morphogen molecules, and the fuzziness of the gradients is significant.

How exactly this statistical variation maps onto the precision of embryonic growth is not yet well understood, but we can already ask about the constraints it places on patterning. In a beautiful 2013 paper,

William Bialek and colleagues at Princeton University connected morphogen measurements and information theory to deduce how many bits of information are encoded in the early fruit fly. If a transcription factor can only be read as "high" or "low," it can be codified by one bit of information—one bit having two states, as we noted in chapter 1. If it can be read as "high," "medium-high," "medium-low," and "low," we would need two bits to encode the four possible states. We don't know how many states the fly's regulatory circuitry can discern for each gene, and so we can't directly calculate the number of bits. Bialek and colleagues realized, however, that the variability of the positions of stripes and edges as one compares individual embryos reflects the number of bits used in the patterning. Essentially, many bits imply high precision and low variability; few bits imply low precision and high variability. Analyzing images of fruit fly embryogenesis, specifically the patterns exhibited by each of the four genes just downstream of bicoid pattern formation, revealed that there are about two bits of information per gene. The four genes together are capable of defining patterns with a spatial precision of about 1%.

Once again, we're presented with an amazing feature of the living world, the generation of wonderful forms by a strikingly small set of instructions. Of course, a maggot may stretch your definition of "wonderful," but if you're unimpressed by fruit flies, keep in mind that these tractable, convenient organisms make possible the discovery and characterization of phenomena that turn out to be widespread. You, too, probably didn't require too many bits at the start.

KNOWING YOUR NEIGHBORS

In the examples of patterning we've seen so far in this chapter, we haven't really had to think about cells. We could imagine fields of morphogens or slabs of tissue without considering the discrete units that respond to or form them. For the early fly embryo, this isn't actually a simplification; the ellipsoidal embryo doesn't consist of discrete cells but rather lots of nuclei floating in a shared cytoplasmic sea. Later, membranes grow to separate the nuclei and form cells. In vertebrate embryos, including humans, distinct cells are present from the start.

In either case, once cells exist they have additional tools at their disposal for patterning.

Cells can convey and receive signals through contact with other cells. We've already seen this with membranes and immune cell signaling, in which membrane-anchored proteins reach across the gap between cells to recognize partners and trigger specific responses. Cell-cell contact plays a major role in development as well, especially in fine-scale patterning. Imagine a layer of cells that can each express genes A and B. These genes might be the defining factors for particular types of cells, A types and B types. Suppose that any cell in contact with an A cell is instructed not to itself express the A gene; it expresses B and becomes a B-type cell. Any cell that isn't touched by an A cell expresses A and becomes an A cell. Given these rules, we'd expect a mosaic that might look something like this honeycomb-like illustration, where each A cell (dark) is surrounded by a ring of B-cell neighbors (light).

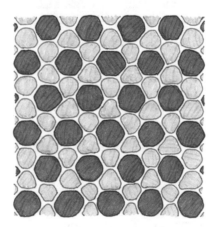

This sort of patterning is commonplace. You can hear, for example, thanks to thousands of hair cells in your inner ear, so called because out of each projects a bundle of membrane-bound pillars, reminiscent of a tiny tuft of hair. Each hair cell is surrounded by a cohort of supporting cells in an arrangement that develops from exactly the patterning described above, referred to as *lateral inhibition*.

Lateral inhibition governs the layout of cells in insects' compound eyes, the specification of smooth muscle cells that make up artery walls, the generation of hormone-secreting cells in the pancreas, and more.

Though predicted since at least the 1970s, this mechanism was first clearly demonstrated in a developing animal in the mid-1980s by Chris Doe and Corey Goodman (the former now a colleague of mine at the University of Oregon) through clever experiments in which they focused a laser to destroy specific cells in fruit flies, eliminating these as neighbors of other cells that, now uninhibited, could express the genes that would turn them into neurons.

How can a cell control its neighbors' fates? For many different cell pairs in many different organisms, the key molecule is a membrane-anchored protein called Notch. Notch passes through the outer membrane of cells that express it. The external segment can attach to targets like Delta, another membrane-spanning protein, extending from a neighboring cell. The intercellular handshake of Delta on one cell grasping Notch on the other triggers a change in Notch's shape that exposes parts that would otherwise be hidden, parts recognized by other proteins that can sever amino acid chains. (It's perhaps disturbing to imagine axe-wielding proteins patrolling the cell, ready to cleave their colleagues in two as soon as they're provoked, but triggered destruction is a recurring theme in biology.) Notch is first cleaved at a site just outside the cell, releasing a large chunk of the external segment that diffuses to take part in other reactions, and then the internal part is liberated from the membrane anchor. This internal segment diffuses within the cell until it ends up in the nucleus, where it binds to other proteins and influences their binding to DNA; in other words, altering the activity of various transcription factors and thereby controlling the expression of various genes. One of the genes that the Notch fragment inhibits is Delta, decreasing its expression, so that a Delta-contacted cell won't be a Delta-expressing cell. Delta, therefore, is the A of our A-B schematic.

Notch and Delta orchestrate lateral inhibition. Notch can, however, have other partners displayed by other cells that trigger different sequences of cleavages and gene regulation. Some can cause the contacted cell to adopt the same fate as the A cell, spreading this type through a tissue. There is a rich variety of patterning-related proteins that can be present at cell-cell junctions, more than just Notch and Delta, as well as a variety of binding strengths, binding rates, and gene response functions that these proteins can show. Like a jigsaw puzzle in which a complex picture can emerge from the piece-by-piece assembly of neighboring tiles, contact-driven signals can orchestrate robust, intricate patterning from simple, local rules. Again, we see self-assembly at work, as cells contain within them the instructions that give rise to their large-scale organization.

KNOWING THE TIME

Groups of cells can organize in time as well as space, and can even use temporal cues to paint spatial patterns. We look at an example in which timekeeping enables the repetition of form. Tiger stripes, centipede legs, and the vertebrae of your spinal column all show regular features, one after the other, not exactly identical but quite similar. Consider the spine. Superficially, its bony regularity is reminiscent of the stripes of the young fruit fly, so we might expect a similar developmental origin from overlapping morphogens. In the early fly, however, the size of the animal is stable while genetic circuits build up patterns of interaction. Your spine, on the other hand, develops from segments that emerge while you are rapidly growing, as is the case for all vertebrates. Starting around the third week after conception in humans, the embryo elongates. Rather than being a smooth tube, the developing body features regularly sized chunks called *somites*, occurring in pairs along its length.

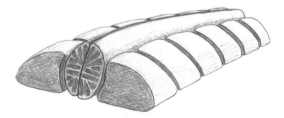

Humans form 42–44 somite pairs, though some fade away as development proceeds. Zebrafish make 30–32, mice about 65, and some snakes more than 400. Vertebrate embryos achieve robust regularity of somite spacing, sustained over dozens or hundreds of somites, by being good timekeepers.

We noted in chapter 4 that cells can use genetic circuits to build oscillators and clocks, with the ebb and flow of gene expression occurring rhythmically in time. Embryos often make use of such clocks. The first few divisions postfertilization are typically synchronized, with each cell in a two-cell embryo dividing at the same time to give four cells, each of the four cells dividing at the same time to give eight, and so on—at least for a little while, until the variety of cells becomes greater and this coordination is abandoned.

The cells of the elongating embryo that form somites also have clocks. Extracted cells in isolation have regular ups and downs of gene expression, and together in tissue these oscillations are synchronized across cells. How can the embryo turn these temporal rhythms into spatial patterns? In 1976, Jonathan Cooke and Erik Christopher Zeeman described an elegant biophysical strategy that subsequent experiments, especially by Olivier Pourquié and colleagues at the Stowers Institute for Medical Research in Kansas City, showed to be the mechanism that you and every other vertebrate animal use for somite generation: connecting a genetic clock to a morphogen gradient.

Imagine an array of cells with gene expression oscillating in sync. The rate of each cell's transcription of some gene, let's say, goes from 0 to 1 to 2 to 3, then back to 0 to begin anew. Depicting these levels as gray, green, blue, and brown, the collection of cells oscillates together, first all gray:

Then all green:

And so on:

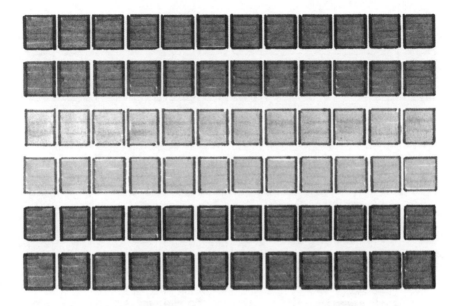

But suppose that each cell's clock has a switch; the timekeeping circuit is active only if the local concentration of some molecule is above a threshold level. If it isn't, the clock stops. Suppose further that the controlling molecule is a morphogen, generated at the tail end of the animal, diffusing and setting up a tail-to-head concentration gradient. The clock will be running in the tail region and will be frozen everywhere forward of some point. As the embryo grows, the tail recedes further from the head, and the location of the threshold morphogen level moves steadily further back as well. The cells will have locked in the levels of gene expression that exist at the moment the threshold concentration passes by—in other words, when the clock stops. If at one location it's 2, at the next it's 3, at the next 0, then 1, 2, 3, and so on, repeating periodically. We can modify our drawing with a pink line, illustrated below, indicating the boundary to the left of which the cellular clocks have stopped; time again proceeds downward, and the tail end of the animal is toward the right. The temporal pattern is frozen in place as a spatial pattern of gene expression.

If high levels of whatever transcription is regulated by the clock generate the somite boundaries, where cells tightly bunch together, and low levels correspond to bulging somite middles, we'll have

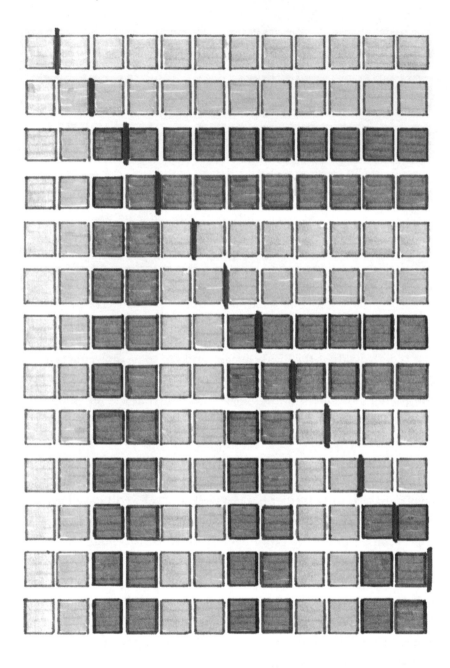

turned our clock oscillations into repeating patterns of physical structure upon which we can construct regular structures, like vertebrae.
What's more, by controlling the speed of the clock, Nature can tune the
spacing of the pattern and the size of the somites. Different animals

have different rates. Snakes make lots of vertebrae by having fast cellular clocks.

We now know a great deal about the specific genes involved in crafting both the clocks and the morphogen gradients. There are several genes at work, especially in the oscillator circuit, some of which encode proteins that bind to the Notch protein we encountered earlier in the context of lateral inhibition. Notch activity rises and falls, and among other things the protein helps maintain the synchrony of adjacent cells. Notch is emblematic of Nature's fondness for using a surprisingly limited set of molecules in a variety of contexts. (A joke already decades old claims there are two types of developmental biologists: those who study Notch signaling and those who don't know they study Notch signaling.)

Gradients and thresholds, contact-driven signals, and stopped clocks aren't the only physical motifs underlying embryonic development. Cells also migrate, stretch, elongate, fold over one another, vary the stickiness of their adhesion, and more, sculpting the living clay of the embryo as it grows. We're still discovering the strategies behind embryogenesis, not only enhancing our appreciation of the transformations each of us underwent from a single cell into a complex animal, but also helping us tackle transformations that don't go as we'd like, as occur in birth defects and cancerous growths.

It's too late to reassure Hans Driesch that biology is not necessarily outside the realm of scientific comprehension, but we can at least reassure each other that the field need not be abandoned. Incomplete though our current understanding is, it's clear that embryogenesis doesn't contradict the laws of physics, but rather is a beautiful manifestation of how physical properties and processes generate living forms.

Organs by Design

Embryos, organs, and every other assembly of cells organize themselves in response to cues patterned in space and time, as we saw in the previous chapter. Groups of cells end up as coherent entities, with distinct biological roles as well as distinct physical properties. Our fat-rich tissue is squishier than muscle, for example, as each of us can readily verify. We've recently realized that these physical properties aren't solely the consequences of tissue and organ formation but can also contribute to their causes. Development influences material characteristics that influence development, in a feedback loop of regulation that further elaborates the tool kit of self-assembly. In this chapter, we look at the roles of physical properties, like softness and stiffness, in guiding collections of cells, and then explore scaffolds for generating, someday, organs outside of bodies.

WHAT DO STEM CELLS SENSE?

Over the course of your life, you'll shed more than a ton of the cells lining your intestine. You won't mind because you're constantly generating new ones. You're also growing new skin cells, blood cells, immune cells, and more. Especially before you were born and for years afterward, you produced many trillions of cells of many different types: liver cells, muscle cells, kidney cells, and so on. All of these were formed from the division of other cells, and proceeding backward along each chain of division and specification, eventually one finds a stem cell. Stem cells are those not yet settled on a fixed identity, instead retaining the capability to generate more than one type of cell, including more stem cells, as their descendants. A single fertilized egg is a stem cell, as it has as its progeny all of an organism's varied cells. In an adult, there are stem cells with more limited potential. For example, one stem cell type generates only the cells of the blood, including red oxygen-carrying cells and immune cells of all flavors. Another variety generates the cells of the intestinal lining, including those that absorb nutrients and those that secrete mucus or digestive enzymes. Given the many possible outcomes, what determines which path a stem cell will take? Will its nonstem daughter cell be, for example, a B cell that grants immunological memory or a macrophage that gobbles debris?

Diffusing molecules form a large part of the answer. As we saw in the last chapter, cells respond to clouds of wandering molecules, tuning gene expression and other activities based on their local concentration. These molecules include hormones, growth factors, and other substances secreted by one cell and recognized by another. Diffusing molecules, however, aren't enough to specify the fates of cells. Recently, we've come to realize that an equally important signal comes from the mechanical and material environment.

Brains are soft, bones are hard, and muscle is in between. Each of these tissues is made up of cells and stuff outside the cell bodies, the latter often in the form of dense meshes of cell-secreted protein. In bone, minerals are incorporated into the protein network, but even prior to mineralization the material is hard, about 10 times as stiff as muscle, which in turn is about 10 times as stiff as brain tissue. The cells and the scaffolds they make influence the rigidity; could the rigidity in turn influence the cells?

In an elegant and influential experiment reported in 2006, Dennis Discher and colleagues at the University of Pennsylvania grew stem cells of a type that can form neurons, muscle progenitors, or bone progenitors on gels of different stiffnesses, with the liquid broth around them kept the same in every case. They found that the stem cells grown on the softest gels, similar in stiffness to brain tissue, transformed into neurons; those on intermediate-stiffness gels into muscle-progenitor cells; and those on the most rigid gels, similar in stiffness to premineralized bone, into bone-progenitor cells.

These newfound identities were evident not only in the cells' shapes—branched for the neurons, elongated for the muscle, and roughly polygonal for the bone formers—but also in the profile of genes that the cells expressed. We already know that self-assembly is amazing; collections of cells are like fabric that sews itself into clothing. But now it seems that the fabric is even more wonderful than we realized: when placed on a soft mattress, it turns itself into a nightgown; on a hard skull, it becomes a helmet.

Mechanics, not just biochemistry, provides cues for determining cell fate. Mechanics also orchestrates many other cellular activities, from the recognition of touch to the perception of sound waves to plants' sensing of gravity to distinguish up and down. The field of mechanobiology, which explores how this mechanical signaling works, has blossomed in the past two decades. Much remains unknown, but some key themes have emerged. One is the importance of channel-forming membrane proteins (chapter 2), whose configuration can be controlled by tension applied to the membrane. Links between the proteins and the internal or external environment can open and close the membrane-spanning gate. Channel proteins can also respond to stresses on the lipid bilayer; a stretched bilayer, for example, can become thinner, and the shorter hydrophobic core (recall chapter 5) can nudge the proteins to adopt an alternative conformation.

Another broad theme is that the network of physical connections between the inside and the outside of a cell can transmit information about forces. Membrane-spanning proteins can adhere, often via various intermediaries, to the external meshwork as well as the internal filamentous scaffolding. Tugging on the cell or the surroundings applies tension to the proteins, which can alter their conformations. This restructuring may, for example, expose sites that were previously hidden, triggering changes in the binding or chemical reactivity of proteins, leading ultimately to the activation or deactivation of transcription factors that orchestrate the activity of genes. Imagine that a stretched protein displays a site to which a repressor protein binds (left illustration); the sequestered repressor won't interact with DNA. In contrast, if the protein is relaxed, the binding site is hidden; the repressor is unbound, free to wander to the cell's DNA and block the expression of its target gene (right). The sketch is extremely simplified compared to the still poorly understood complexity of a real cellular response, but it gets at its essence.

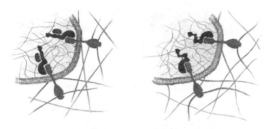

The tugging of proteins takes place even if everything appears motionless. The cell's internal machinery is never still; motor proteins (chapter 2) move, filaments grow and shrink, and the cell itself is constantly pulling. The external mesh isn't active, but its stiffness determines the equal and opposite force it exerts and thus sets the tension experienced by the structurally malleable tension-sensing intermediate.

Mechanical inputs and the material characteristics of the living environment form part of the regulatory circuitry of life and are integrated into the decisions that self-assembling cells make. The details are challenging to determine, but recent work is beginning to unravel the processes at play. Consider your skin, a layered tissue that is constantly losing cells at the outside and replenishing them due to stem cells at the innermost depths. When stretched over sustained periods of time, skin expands, growing extra cells and thereby making more skin. It's a useful response not only for the normal trials skin faces but also for accommodating reconstructive surgeries. To explore how this works, the groups of Benjamin Simons at the University of Cambridge and Cédric Blanpain at the Université Libre de Bruxelles, Belgium, examined mice in which an expanding gel had been placed under the skin. The stretching, the researchers found, led to heightened expression of genes encoding proteins involved in cellular adhesion and the cell's motor protein and filament network. Moreover, stretching induced more stem cell division as well as a greater fraction of the daughter cells being stem cells, ready to generate still more skin. The connection between stretching and cell fate decisions involved specific transcription factors that the researchers were able to identify; engineering mice to lack these transcription factors prevented the stretch-induced stem cell responses. How these particular transcription factors are coupled to the cellular scaffolding remains unknown, but at least the pieces of the puzzle are beginning to be discovered, and one can hope that further elaboration would be of use, for example, in designing therapies that call for accelerated skin regeneration.

Stiffness isn't the only material characteristic cells sense to guide their development. We're creatures of fluids as well as solids. Blood courses through our arteries and veins, and it turns out that fluid flow can induce stem cells to transform into the cell types that line blood vessels. All of our tissues, organs, and internal spaces have their own rigidity, viscosity, elasticity, and other material features whose development is intertwined with that of the cells that make them. While we try to make further sense of these connections, we push forward with ways to engineer multicellular structures, aided by our rapidly expanding insights into the role of the physical environment in guiding organ development.

ORGANS ON A CHIP

If you need a new heart, why not grow one? The dream of organs in a vat, guiding themselves into their proper forms like fruits in a garden, holds obvious appeal. Imagine replacing an injured eye with a fresh one, perhaps seeded by your own body's cells, or substituting a smashed finger with true flesh and bone rather than an inorganic prosthetic. Beyond repairing damage, assembly of isolated organs would enable the study of their development and the testing of drugs free of the complications, both practical and ethical, of studying organs inside of animals. We're still far from this vision, but our progress is accelerating dramatically, especially in the context of self-assembled cell clusters called *organoids* and partially human-assembled "organs on a chip."

We've grown animal (and plant and fungal) cells in the lab for many decades. Much of what we know about cellular biology—the filamentous network of internal scaffolding, the trafficking of cargo, and more—comes from such "cultured" cells. These tend, however, to be essentially two-dimensional: cells spread out on a petri dish, slab of gel, or other flat surface, bathed by a nutrient-rich broth.

There are obvious limitations to this approach. A layer of heart muscle cells can rhythmically pull and push, but it can't form heart-like tubes and chambers. This is more than just the trivial consequence of a flat geometry. Recall from the last chapter that cellular decisions often depend on the arrangement of contacts with neighbors and the shapes of morphogen gradients, both different in two versus three dimensions. The molecular, chemical, and mechanical cues of the three-dimensional environment are critical factors in the development of an organ.

The artificiality of two-dimensional cell clusters has been appreciated for over a century, and the earliest efforts to transcend it are nearly as old. Ross Harrison, whom we briefly encountered in the previous chapter, reported in 1906 on the growth of nerve fibers into a clotted drop of lymph fluid, seeded by a bit of embryonic frog tissue. In subsequent decades, several research groups showed that embryos of various species could be split apart into discrete cells that, if free in three dimensions, could coalesce into aggregates that recapitulate some aspects of normal embryonic form.

Over time, researchers realized that the meshwork of proteins outside of cells, called the *extracellular matrix*, is crucial to cell function, not just forming the scaffolding for tissues but also providing the mechanical and chemical cues that guide gene expression and even cell fate. In the 1980s, for example, Mina Bissell's group at Lawrence

Berkeley National Laboratory in California grew mammary gland tissues capable of secreting milk under the guidance of the appropriate matrix material, a remarkable demonstration that these collections of cells not only looked as they should but behaved as they should. Many other tissues, including cancerous growths, began to be nurtured using an expanding array of three-dimensional culture techniques. The field has taken off in the twenty-first century as our understanding of the underlying mechanisms intersects our understanding of the ideal seeds for tissues in a vat: stem cells.

Combining stem cell techniques and three-dimensional culturing methods gives us an amazing repertoire of functional, self-generating cellular assemblies, tantalizingly close to "organs in a vat," of almost any type. Such assemblies, whether stem-cell-derived or not, are referred to as *organoids*. The cells sloughed off at the surface of your intestine are replaced by the offspring of intestinal stem cells situated near the bottoms of billions of pocket-like pits. In 2009, Hans Clevers's group in Utrecht, the Netherlands, showed that a *single* intestinal stem cell, appropriately raised in a three-dimensional matrix, can divide into a community of cells shaped like a bumpy ball, with a well-defined interior space surrounded by the same sorts of cell surfaces as the intestinal interior, and with stem cells (blue in the illustration) near the bases of small pockets.

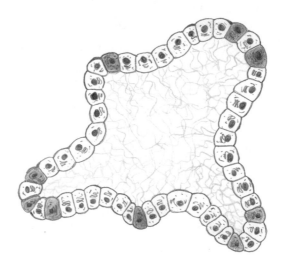

In other words, the intestinal stem cell generates an intestine-like organoid, similar enough in form and behavior to allow, for example, its use in studies of drugs targeting intestinal diseases.

Invoking the eye, Yoshiki Sasai and colleagues at the RIKEN Institute in Japan grew organoids that shaped themselves not into balls or shells but rather into the roughly hemispherical curve of the early "optic cup" (the back of the eye), initiated by stem cells of the sort that transform themselves into retina cells.

A few years earlier, in 2008, the same group demonstrated that mouse stem cells could be grown into balls of connected neurons similar in structure to the cortex region of the mouse brain. In 2013, the lab of Juergen Knoblich at the Austrian Academy of Sciences in Vienna built "cerebral organoids" that recapitulate several of the layers and structures of a normal brain, with working neurons and areas resembling the nascent prefrontal cortex, hippocampus, and more. Though they were far from functioning brains, the organoids were immediately useful. Knoblich's group investigated the puzzling origins of microcephaly, a developmental disorder resulting in a small brain, by using stem cells from a human patient as the seed for the organoid. In contrast to those derived from normal individuals, the patient-derived organoids showed fewer rounds of replication by a certain class of stem cells, driving an overall shortage of cells. Though the possibility of cerebral organoids developing sensation or consciousness is very far off, scientists and philosophers are already collaborating to map the ethical issues involved, including the question of how to assess and interpret the capabilities of a collection of neural cells.

Guts, eyes, brains—the full list of organoids developed to date is much larger than this, and it continues to expand. As noted, organoids are powerful tools with which to study development, disease, and drugs in fundamentally new ways, providing organ-like objects not ensconced in an animal body, and potentially derived from human cells. It's not much of a stretch to imagine that with further technological developments they may progress from being organoids to full-fledged organs, ready for transplantation into human recipients.

Beyond their practical utility, it's worth marveling at the biophysical lesson organoids convey. We've commented several times on the theme of self-assembly, for example, at the scale of molecules with proteins folding themselves into specific shapes and at the scale of organisms where entire bodies are generated from intrinsic rules. Here we see self-assembly being employed in a modular way at scales in between: the components of individual organs carry the instructions for their own organization. It's as if we not only have a car that forms itself from a small seed, but that a scrap of engine, bathed in the appropriate motor oil and held by the appropriate clamp, grows into a complete, growling engine or a scrap of the driver's seat, held gently, grows into a new seat; each piece on its own is, to some extent, self-sufficient. Nature harnesses self-assembly through a cascade of scales.

The self-assembly of organoids is assisted by our design of the appropriate extracellular matrix, though once set this is left alone. What if we were more hands-on, more deliberate in the operation of machinery outside the cells? Building small things, unconnected to biological applications, has been one of the great triumphs of the past half century of human civilization. The chips performing calculations inside mobile phones, for example, each contain billions of transistors squeezed together in a square inch of area, churned out by factories with astonishing speed and reliability. Our microfabrication abilities extend beyond electrical components like diodes and transistors. With materials like plastics and gels, we can fashion small channels, junctions, valves, and pumps at sub-millimeter scales—just the sorts of things one would need to bring nutrients or stimuli to groups of cells.

Merging microfabrication and cell culture gives us "organs on a chip," of which, as with organoids, there is a dazzling and expanding variety. The lab of Donald Ingber at the Wyss Institute for Biologically Inspired Engineering at Harvard University has pioneered several of these, including a "lung on a chip" in 2010. In this design, a porous, soft, very thin silicone sheet separates two chambers—one filled with air, the other filled with a blood-like, watery solution. As we discuss further in chapter 11, lungs are fundamentally an interface between air and water at which gases are exchanged. On one side of the sheet, the researchers cultured cells of the type that line the air side of the

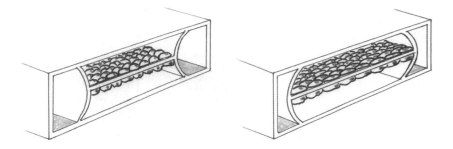

lung interface; on the other, cells that line the blood vessels. The clever part is that, at the edges of the sheet, separated by thin walls, are chambers that can be filled with air or evacuated. If under pressure, they compress the sheet and the attached layers of cells; if under vacuum, they stretch the sheet and the cells.

We therefore have a device that mimics not only the structure of the lungs but also their dynamics, stretching and contracting with, if we wish, the same rhythms as natural breathing. The chambers, sheets, and valves are all constructed by microfabrication methods that can cover a chip with a mosaic of pseudolungs, all amenable to imaging with a microscope. Ingber and colleagues showed that the uptake of particulates across the cellular boundary, a concern for both air pollution and drug delivery, is enhanced by the periodic mechanical pulsing of the cell sheets. Since then, scientists have constructed hearts on a chip, kidneys on a chip, stomachs on a chip, skin on a chip, and more, even stringing multiple organs together to build the hyperbolically named but nonetheless impressive "bodies on a chip." Much of the cell culture in these systems is still two-dimensional; integrating the three-dimensional self-assembly of stem-cell-derived organoids with the fluid handling and mechanical scaffolding of organs on a chip is an exciting area for ongoing work.

The semiartificial constructions of lab-cultured stem cells, organoids, and organs on a chip all provide us with handles on the awe-inspiring phenomena of multicellular organization. So far, however, we've treated the cells of an organism as all belonging to the same species. In the next chapter, we see that this isn't quite true—you contain multitudes of microbes—and we explore some aspects of these animal-associated microbial communities.

The Ecosystem inside You

You likely think of yourself as human. Your body is made up of a few trillion human cells, each enclosing a human genome, lending support to your concept of species identity. However, your body is also home to several trillion microorganisms—mostly bacteria, with some archaea and eukaryotic microbes as well—so many that if you held a vote, your human cells would probably lose. These microbes inhabit your mouth, your skin, and every warm, wet surface you can imagine, but by far the largest fraction resides in your intestines. This isn't a peculiarity of humans. All animals are hosts to a large and diverse assembly of gut microbes, and without these fellow travelers we'd have great difficulty functioning. Plants, too, have microbial partners, especially associated with their roots.

We've known of the existence of intestinal communities, often referred to as the gut microbiota or the gut microbiome, for well over a century. Our interest in them and our awareness of their importance, however, have exploded over the past two decades, driven by the technological revolution of DNA sequencing. Studying bacteria prior to this typically required growing them in a laboratory culture. Unfortunately, most bacteria stubbornly refuse to cooperate. Some, like those normally at home in the human intestine, find the amount of oxygen in the ambient atmosphere toxic. Some require especially acidic or basic conditions. Some need exotic nutrients, perhaps produced by other microbes. It's possible to meet these conditions, but it's often difficult, and moreover the solution may be different for each member of the community of interest. Therefore, for most of the time that we've been aware of gut microbes, we've known little about them.

DNA sequencing changed all this. We explore how sequencing works in part III. It suffices for now to recall from chapter 1 that we can make abundant copies of any DNA that we find. We can feed this material into a machine that reads it, returning the sequence of As, Cs, Gs, and Ts that make up the DNA fragments. This methodology can be applied to DNA from all sorts of sources, and it's radically revamped our notions of ecological diversity. In a pioneering study in 2004, a group led by Craig Venter, one of the key figures responsible for inventing modern DNA sequencing technology, scooped a few hundred liters of water from the Sargasso Sea, pureed the contents, purified and amplified the DNA (as in chapter 1), and discovered a million previously unknown genes from hundreds of novel bacteria. We've now similarly explored through sequencing environments from soil to subway stations, tips of tongues to fecal samples. (The fecal samples give us a snapshot, albeit an indirect one, of the microbes present in the intestine.)

In all these habitats we find teeming ecosystems, rich in cells and species. As we did for organs, tissues, and embryos, we can ask whether biophysical principles can help us make sense of these ensembles. In chapter 6, for example, we wondered why bacteria swim, and found an explanation in navigation to "greener pastures" of nutrients. In the tumultuous landscape of the gut, we can wonder whether similar strategies apply, or if bacteria have other motivations for moving. We can investigate whether groups of microbes self-assemble, either tangibly into physical structures or more abstractly into networks linked by biochemical exchanges. We can also apply the biophysical tools we've developed to probe the workings of the microbial ecosystem, for example, manipulating the regulatory circuits that guide bacterial decision making and observing the consequences. More than in prior chapters, we visit the frontiers of our understanding, at which answers, and even questions themselves, are still coalescing.

CATALOGING DNA

Before returning to the intestinal microbiome, I'll say a bit about two common methods to take a census based on DNA sequencing. The first makes use of a bacterial gene called the 16S ribosomal RNA (or 16S rRNA) gene. What it does isn't important here; what matters is that every species of bacterium has this gene and that some regions of the gene are identical across every species and some are different. The identical regions (gray in the illustration), when translated into RNA, correspond to sections that are crucial for the three-dimensional shape of the RNA molecule. The nonconstant regions (colored) are the record of billions of years of evolutionary variation, as different species adapted the basic rRNA architecture for slightly different purposes.

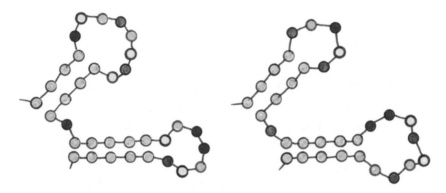

The consequence of all this is that we can conveniently use the same primer sequence (see chapter 1), corresponding to one or more of the identical regions, as the starting point for amplifying the DNA from any bacterium, getting countless copies of all the 16S rRNA genes present in our sample. Thanks to the variable regions, the complete rRNA genes are different enough that sequencing these DNA copies reveals the distinct signature of each contributing species. The 16S rRNA gene, therefore, is like a handle and a fingerprint all in one.

The drawback of 16S rRNA sequencing is that it reveals the identity of the bacteria present but nothing else. It's like having a list of names of all the people in a town, but no information about their ages, occupations, incomes, interests, or anything else that might allow you to assess what the town is like. If the 16S rRNA sequence matches that of some known bacterium, we can latch onto that information; but since most bacteria are unknown, this isn't often the case. Furthermore, closely related bacterial strains might have indistinguishable 16S sequences; it's like our list of names is a list solely of last names, not separately identifying, for example, siblings in the same family.

An alternative approach is known as shotgun sequencing. Here we amplify, break into sequence-able pieces, and sequence every scrap of DNA we collect, afterward computationally assembling them into genomes. Imagine, for example, that you had the fragments of sentences that were written on strips of paper, with multiple copies made of each sentence. Some read "a thousand times before his death"; some read "good or bad, but thinking makes it so"; some read "The fault, dear Brutus, is not in our stars"; some read "There is nothing either good or"; some read "A coward dies a thousand times"; some read "not in our stars, but in ourselves." Even if you didn't know anything about grammar, syntax, or Shakespeare's plays, you'd deduce from the overlaps ("good or," not in our stars," "a thousand times") which fragments came from the same source sentence, and you could assemble them into three distinct quotes. Similarly, we can write computer programs to determine the optimal overlap and alignment of potentially billions of DNA sequence fragments, reconstructing the full genomes from which they came. This is more complex and costlier than 16S sequencing, but it tells us what genes the members of our sample of interest contain, and therefore what proteins they can make and, in principle, what activities they can perform.

THE GUT MICROBIOME AND YOU

Thus, from the undignified starting material of a stool sample, we can glean the makeup of the ecosystem contained within you. Even from the earliest investigations, several striking features were evident. These microbial communities are very diverse; you harbor hundreds of different bacterial species. These species are special, not simply mirroring the stowaways on the food you eat or even the inhabitants of your mouth, but rather forming a group specifically acclimated to the intestine. Your gut community is unique but not wholly so; there's a lot of overlap with the communities of other people, which increases as we focus on the same geographic region and especially the same home. There's also considerable, but not perfect, overlap between the set of species in you now and the set inside you a few months ago;

some gut bacteria are itinerant travelers, but many stick with you for the long haul.

The aspect of the intestinal microbiome that attracts the most attention, however, and that has made "gut bacteria" a household phrase, the subject of newspaper articles and advertisements, is the correlation between the composition of the gut microbiota and a wide range of complex diseases. Disorders as varied as diabetes, inflammatory bowel disease, gastrointestinal cancers, and even neurological ailments, such as multiple sclerosis and Parkinson's disease, all seem to be associated with altered makeup of the microbial communities of their sufferer's guts, compared to healthy people. These differences aren't characterized by the presence or absence of just one or two species. Unlike "classic" diseases like tuberculosis (caused by the bacterium *Mycobacterium tuberculosis*) or bubonic plague (caused by the bacterium *Yersinia pestis*), it seems that a broad shift in the abundance of tens or hundreds of species somehow takes place in the course of these more mysterious diseases.

Of course, correlation isn't causation. It's difficult to determine whether aberrant microbial community structure is a symptom of disease or whether it contributes to its origin, and the two options are not mutually exclusive. Still, the potential for a causal arrow from microbial ensembles to devastating diseases spurs an intense amount of effort aimed at elaborating these links and, one hopes, deliberately curating the intestinal ecosystem to foster health and fight disease.

At least for some disorders, there does seem to be an arrow one can draw from gut bacteria to health. Pernicious, recurrent infections by the often toxic bacterium *Clostridium difficile*, for example, have proved amenable to treatment with fecal microbiota transplantation. As the name implies, the procedure involves transferring fecal matter from a healthy donor to the patient, overwhelming the patient's aberrant native community with the new microbial immigrants. Fecal microbiota transplantation also shows tantalizing efficacy for inflammatory bowel disease and other disorders, though the results to date are highly variable. It's deliberate that the word *transplant* in the procedure's name echoes organ transplantation. The gut microbiome is, in a sense, an organ, serving a physiological role in its host despite not being composed of the host's own cells.

These issues of health and disease raise many questions. Could we replace the "fecal" part of fecal transplantation with a pill containing a carefully chosen set of bacteria, both to reduce the odds of unintentionally infecting a patient with harmful microbes via the transplant and to make the transplantation less aesthetically unpleasant? Would the pill even require bacteria at all, or simply nutrients preferred by the microbes we'd like to boost? (These questions spur a whole industry, that of probiotics.) What are the bacteria actually doing that affects, or is affected by, the health of their host? How can we perturb or even perhaps "reboot" the gut microbiome?

CAN WE UNDERSTAND THE GUT MICROBIOME?

We don't know the answers to these questions. The field of microbiome research is thrilling and vibrant. It is also a mess. Contradictory, inconclusive, and overhyped studies abound, along with many gems. Rather like the Wild West, adventurers and fortune seekers both good and bad rush in, and laws and sense come later. Many factors contribute to this chaos. First, as mentioned, the gut microbiome is highly variable. Tolstoy's assertion that "all happy families are alike; each unhappy family is unhappy in its own way" applies fairly well to conventional organs. Healthy hearts, for example, all resemble one another in their anatomical structure and in their rhythmic contractions. In fact, we can use structural aberrations or arrhythmic electrical signals as reliable indicators of disease. The gut microbiome, in contrast, doesn't have a stereotypical membership list. Different healthy people have quite different gut microbiomes, leaving us with the challenge of discerning the subtle statistical features that distinguish health and disease.

Second, fecal samples, as useful as they are, are an indirect reporter of what's present in the gut. Strictly speaking, they're an indicator of what is *not* in the gut, and it's a rather strong assumption to consider fecal samples as representative of what's left behind. In a tour de force experiment, a team of researchers led by Eran Segal and Eran Elinav at the Weizmann Institute of Science in Israel compared the microbiome of fecal samples with that obtained by more invasive, direct sampling of the intestine, finding that the former is clearly not the same as the latter, in both humans and mice. In the same study, the scientists fed mice and people supplements containing commonly used, commercially sold, probiotic bacteria, and found quite limited and highly variable colonization of the intestine by these "desirable" microbes. Examining what goes into or out of the gut is relatively easy, but it may not tell us enough about what's actually within the gut to make sense of the intestinal microbiome.

Third, it's difficult to do controlled experiments that untangle cause and effect, or that clearly distinguish meaningful changes from random variation. Every food we consume is a source of nutrients that affect some microbes differently than others; every room we step into is a potential source of new microbes. These incessant perturbations are hard to get rid of. We can't raise groups of people that are identical in every respect except for their intestinal microbes and see how they progress toward different diseases and disorders. We can't easily compare existing groups, for example, people from different regions, and be sure that differences in health are attributable to their gut microbes and not to many other confounding variables.

With animals, we can achieve a much greater degree of control. Mice, zebrafish, and other animals can be raised "germ-free," devoid of any microbes, and then either maintained in this pristine state or introduced to particular microbial species. Such studies have revealed that germ-free animals exhibit a wide range of abnormalities; resident microbes seem crucial to the training and stimulation of immune cells, the proliferation of cells that line the gut, and more. The control afforded by these experiments has in several cases enabled discovery of specific chemical factors by which microbes influence their hosts. The lab of my colleague at the University of Oregon, Karen Guillemin, who has pioneered germ-free zebrafish studies, found that germ-free fish larvae have a paucity of the insulin-producing beta cells of the pancreas, a defect that can be reversed by bacterial colonization or by the application of a specific protein that certain gut-native bacteria secrete. In humans, type 1 diabetes is characterized by destruction of beta cells; the zebrafish finding may point to previously unimagined paths toward beta cell regeneration. Microbiome-assisted development has been observed for several other organs and tissues, for example, bone growth in infant mice. Gut bacteria, it seems, have figured out a variety of communication pathways that link them to their host. Conversely, the animal host listens to its microbes, especially for inputs into how it should be developing early in life. In a sense this is surprising; it seems risky for animal development to depend on nonanimal partners, especially flighty and variable creatures like bacteria. On the other hand, animals evolved in a world already occupied by microbes. Their presence is unavoidable, so relying on microbes may have been nearly as risk-free as relying on the laws of physics. This last statement is a controversial one—I'm not even sure I believe it—but it's fair to say that our concepts of animal development are being altered by our expanding insights into microbial communities.

Returning to the pessimism of a few paragraphs ago, however, I will point out that raising germ-free animals is quite difficult. Mice can be kept germ-free to adulthood, but it requires considerable effort and expense. Zebrafish are easier, but still not easy; my lab and the labs of my colleagues struggle with this routinely, as bacteria and fungi exploit every possible chance to leap into a germ-free fish. Moreover, zebrafish can't be kept germ-free to adulthood (at least, not yet), because they need live food for adequate nutrition, food that brings with it its own microbes. As a result, a remarkable number of much-touted studies, especially in mice, are based on analyses of single-digit numbers of animals. It's a bit like trying to predict the outcome of a national election by polling a single-digit number of voters—the variability and complexity of the system demand much wider sampling.

For all the reasons we've seen and more, many microbiome-related findings are not as robust as one might hope. Links between human obesity and the composition of the gut microbiome, for example, have faded in strength in the decade since their first announcements. This isn't to say that such links don't exist—gut microbes certainly participate in many digestive processes and influence functions such as fat absorption that are tightly connected to obesity—but the roles of varied microbes aren't as simple as one might hope. Similar pitfalls occur for other features of health and disease. It may seem strange that science is prone to such missteps. Among many scientists, however, these issues of reproducibility are under increasing scrutiny in many fields.

THE GUT MICROBIOME AND ME

My goal here, however, is not an exhaustive survey of what we do and don't know about the gut microbiome, nor is it to examine the structural problems of contemporary science, fascinating though both topics are. Rather, it is to ask whether a biophysical perspective can help us understand the gut microbiome. We wonder, for example, whether the architecture of bacterial colonies can be illuminated by concepts of self-assembly, and whether general strategies for bacterial navigation emerge amid the confined, churning landscape of the intestine.

Even more than for the other topics in this book, the answers to the questions we wish to ask are unknown; making sense of our intestinal ecosystem is very much a work in progress. This quest is, in fact, the major focus of my research lab at the University of Oregon; so in addition to commenting on what some general principles may be, I'll describe how I dropped nearly all my other research to pursue the idea that there may be physics in the strange substance of the gut microbiome.

My decision was admittedly odd. I'm a physicist by training and a professor in a physics department. As we've seen, physics is more than magnets and quarks and lasers. There's a lot of physics in the living world, but even among biophysicists it's not obvious why one might consider physics relevant to the messiness of guts, in contrast to the precise choreography of protein folding or the mechanical rules of DNA packaging. Why would I, and a small but growing number of other biophysicists, gamble on the biophysics of the microbiome being something worth exploring?

Imagine a tropical forest, dense with plants and animals. If you knew that there exist organisms called monkeys, leopards, elephants, and trees, but were somehow unaware that trees are stationary, that monkeys climb trees but elephants don't, that leopards hunt monkeys but leave elephants alone—if you were unaware of any aspects of the behavior, locations, size, and mobility of the forest's organisms, it would be hopeless to try to construct an accurate picture of how the forest ecosystem works. If you were unaware that on a rocky shore the tide comes in twice a day, that starfish are mobile predators, and that sea lions swim in to snack, you'd similarly struggle to understand the tidal zone ecosystem, no matter how much of the creatures' DNA you swabbed off the rocks or scooped up from the seawater. This seems obvious for macroscopic ecosystems, and the lesson that structure and dynamics matter is a very general one.

As mentioned, however, most of our information about the gut microbiome comes from DNA sequencing methods that are blind to its layout and activity. I became increasingly fixated on this lack of biophysical understanding about a decade ago, coincident with a global explosion of interest in the intestinal microbiota and many conversations with my aforementioned colleague Karen Guillemin. Around the same time, I became entranced by developments in microscopy, especially a method called *light sheet fluorescence microscopy*, which makes possible fast, three-dimensional imaging of large fields of view ("large," by microscopy standards, being many tenths of a millimeter). My research group, therefore, built its own light sheet fluorescence microscope and pointed it at the intestines of live zebrafish larvae, realizing that we could capture images and movies that would span the entire gut, yet be precise enough to see single bacteria. No one had done this sort of imaging before in any vertebrate animal. We took advantage of the optical clarity of the larvae and their amenability to germ-free derivation, exposing fish devoid of any intestinal microbes to just one or two species to see what, in the absence of confounding complexity, the organization and behavior of these microbes would be.

If we were unlucky, I reasoned, we'd find featureless swarms of bacteria, the same regardless of species, and the same as what we'd see in a beaker or a test tube. We'd perhaps occupy ourselves with measurements of gut bacterial growth rates or some such boring but possibly useful tasks.

Thankfully, Nature was kind to us. Even from our first observations, it was evident that a wonderful variety of forms are present. Some bacteria swim freely; some clump together. Some prefer the forward part of the gut; some are mostly behind. There are clear candidates for physical features that could influence how an intestinal ecosystem works and that might help us figure out how to tinker with competition and cooperation among gut microbial species.

Nearly all my lab's explorations with zebrafish larvae involve native bacteria, found in the zebrafish gut. However, for the first vignette that illustrates the physical backdrop to microbiome dynamics, I'll describe an experiment using a nonnative species, *Vibrio cholerae*, the bacterium that causes cholera. *Vibrio cholerae* has been studied intensely for over a century, and while cholera is not the globally devastating scourge it was a hundred years ago, it still kills around 100,000 people each year. I had barely thought about cholera until an unusual meeting at Biosphere 2, the site of an ill-fated attempt in the early 1990s at operating a sealed, self-contained experimental ecosystem. There I met Brian Hammer, a microbiologist at the Georgia Institute of Technology, and Joao Xavier, a microbiome expert at Memorial Sloan Kettering Cancer Center in New York City. We weren't locked into a concrete building and forced to grow our own food, but rather were attending a workshop organized by the Research Corporation for Science Advancement, a small private funding agency. (Biosphere 2 is now a tourist attraction and meeting site administered by the University of Arizona, which also runs experiments in the project's glass-domed and no-longer-sealed buildings.) Through conversations, we realized that the means by which *Vibrio cholerae* invades a human intestine are not at all well understood. It's not an empty space the bacterium wanders into, but rather a gut densely packed with trillions of resident microbes amid which *Vibrio cholerae* must somehow gain a foothold.

For years, Brian has been studying an amazing tool that *Vibrio cholerae* and many other microbes have, a syringe-like device called the type VI secretion system with which bacteria stab adjacent cells and inject toxins. We wondered whether *Vibrio cholerae* might be using this system to help invade the gut, and whether our combination of light sheet microscopy and zebrafish larvae would enable us to assess this. In addition, no one had ever watched *Vibrio cholerae* colonize and compete within a live animal gut; who knew what thrilling things we might find? Brian's group engineered several strains of *Vibrio cholerae*, including one variant in which the type VI secretion system genes were always on, so the bacteria were always ready to stab (left illustration), and one variant in which the syringe apparatus was defective, so the bacteria were incapable of stabbing (right). Joao's group conducted petri dish experiments to examine and visualize bacterial-bacterial killing, itself quite beautiful to watch. We looked inside the fish.

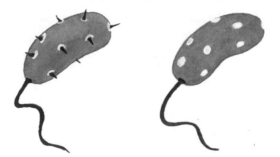

As the simplest possible intestinal invasion experiment, we precolonized initially germ-free fish with a single native bacterial species, and after 24 hours added to the surrounding water one of the *Vibrio cholerae* strains. When potentially invaded by the stabbing-defective *Vibrio*

cholerae, the native bacteria looked just like they do when on their own, abundant and forming dense colonies. In contrast, the natives were annihilated after introduction of the always-stabbing invader; within a day, the populations would drop by more than a factor of 100 on average, often disappearing completely. By imaging, we could watch bacterial colonies losing their grip, steadily receding down the intestine until eventually expelled from the fish. At this point, we were excited to see the type VI machinery making a clear difference, and we assumed that the armed *Vibrio cholerae* were killing the native bacteria, dislodging them. We kept looking, however, and turned our attention to what the zebrafish itself was doing. Just like your intestines, the larval zebrafish intestine pulses periodically, squeezing to churn and move its contents. Savannah Logan, the graduate student leading the experiments, observed that the intestines of fish colonized by the always-stabbing *Vibrio cholerae* showed much stronger contractions than germ-free fish, or fish colonized by the other strains. Analyzing the images, the magnitude of the contractions was about 100% larger in fish harboring bacteria with the active type VI machinery. It looked, therefore, like bacteria weren't stabbing their competitors, but stabbing their host.

Proving this required more genetic engineering and more microscopy. A part of one of the proteins that constitute the bacterial syringe was already known to be toxic to eukaryotic microorganisms, like amoebas, via disruption of the filamentous scaffolding present in all eukaryotic cells but absent in bacteria. Brian's group engineered yet another *Vibrio cholerae* strain that lacked this piece but that formed an otherwise functional syringe. Colonizing fish and performing the same invasion assay, we found normal levels of gut contractions and normal, robust levels of the native gut bacterial species. *Vibrio cholerae*, therefore, was not defeating its bacterial competitors by killing them but rather by using its syringe to poke its host, antagonizing the fish to strengthen its intestinal contractions in response and thereby expel aggregated resident microbes. Conveniently, *Vibrio cholerae* itself was unaggregated, motile, and individualistic. This marked the first finding that any bacteria could use their syringe machinery to manipulate an-

imal physiology. More broadly, it highlighted that the physical landscape of the gut is crucial for governing the gut microbiome, through
mechanisms that are fundamentally unknowable if one looks solely at
DNA sequences or test tube experiments.

Will our findings help cure cholera? I doubt it. Of course, the traditional answer, and the requisite answer in any news report or press release, is to say yes, or at least imply that cures for every even tangentially related ailment are just around the corner. For cholera, however,
there's a more important issue than the long and unpredictable chain
that connects basic laboratory science to practical treatments: cholera
is already easy to cure. The treatment, except in the most severe cases,
is water with salts and sugars. The shockingly large number of people
who die each year of the disease is a sad testament to the inadequacy
of sanitation and public health systems throughout the world. Why,
then, should we care about *Vibrio cholerae's* type VI secretion system?
Aside from its intrinsic interest, what excites me is that *Vibrio cholerae*
is just one of many bacterial species with this machinery. Tens or even
hundreds of species in your intestinal microbiome have a type VI secretion system, and so understanding its role in the gut may help us
understand what determines your microbiome composition. Manipulating the type VI secretion system in a range of bacteria might give us
a long-sought path to altering the gut microbiome, reshaping it to foster
health.

Nearly everything we've looked at in the zebrafish gut has revealed
a strong biophysical signature—some way in which the physical aspects
of behavior or response, whether swimming and navigating, the formation of three-dimensional colonies, or the manipulation of intestinal
forces, are a major determinant of outcomes. As another example, we
found that weak doses of a common antibiotic can induce normally motile bacteria to elongate and entangle and normally aggregated bacteria to form fewer but larger clusters, in both cases leading to severe
drops in the intestinal population as the overly cohesive microbes are
pushed around by intestinal forces. We suspect that this could be a
mechanism behind large and mysterious antibiotic-induced changes in
the human gut microbiome, uncovered through DNA-sequencing-based

methods, which is especially a concern because low levels of antibiotics are commonly found as environmental contaminants. This project, like many of ours, was a collaborative effort between my lab and that of our close colleague Karen Guillemin, and was executed primarily by Brandon Schlomann, a PhD student in physics, and Travis Wiles, a post-doctoral researcher in biology, both happily blurring boundaries between subjects.

Watching bacterial behaviors is great; controlling them could be even better. Our general theme of regulatory circuits resurfaces. In chapter 4, we encountered tools that could activate or deactivate specific gene circuits; recall the color-changing mice. The aforementioned Travis Wiles engineered such handles into the genome of a zebrafish-native gut bacterial species, allowing control of swimming and chemical sensing. These swimming bacteria move through fluid by rotating a corkscrew-shaped, taillike flagellum that extends from one end of the cell. The flagellum and its motor are formed by the self-assembly of many different proteins, including a pair called PomA and PomB (green in the illustration of the base of the flagellum) that form part of the motor. Without PomA and PomB, the flagellum forms normally, but the motor can't generate any torque with which to rotate the flagellum and propel itself. A switch, therefore, that in response to an external chemical cue turns off the *pomA* and *pomB* genes in a bacterium that normally expresses them, or that turns on the genes in a bacterium in which they're normally silent, allows us to control whether these microbes swim or don't swim in the gut. (In our implementation, the external cue must always be present, like a button that

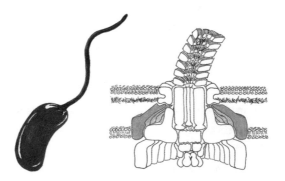

must always be held down for a light to be on. It's not a memory-enabling switch, therefore, but what an engineer would call a momentary switch.)

A switch offers more insights than a simple deletion or a constant activation of a gene. If we delete genes involved in motility, for example, and do not find these bacteria in the gut, it could be because swimming is necessary to persist in the intestine, or it could be because swimming is necessary to reach the fish and colonize it in the first place. Turning off swimming or other behaviors after bacteria have colonized allows us to delineate roles in the specific context of living in the gut. Turning off motility, we found, led to large drops in the populations of these microbes, as they were helpless against intestinal flows that transported them out of the fish and couldn't grow rapidly enough to counteract their losses. More surprisingly, the animal itself could sense these behavioral changes. Using zebrafish engineered to produce green fluorescent protein whenever genes in an immune circuit turned on, we saw large immune responses when the fish was colonized by the normal, motile bacteria, as we expected from earlier observations, but very low responses when the bacteria couldn't swim. The textbook picture of immune cells simply binding to proteins on the bacterial surface can't explain this; the external appearance of the bacteria is unchanged. Motion matters, we suspect, by allowing bacteria to push closer to the boundaries of the gut and make contact with sensory cells. This remains to be proved, but regardless we suspect that the behaviors of bacteria in a gut are as important as the behaviors of animals in a forest for governing the activity of the ecosystem.

We're not the only people excited by the potential of genetic circuit engineering. Many researchers have realized that microbes with memories could record intestinal conditions, with their state after they pass through the gut indicating whether they were exposed to particular toxins, nutrients, or other chemicals along the way. Coupling genetic switches to circuits that make specific biochemical agents could enable the delivery of therapeutic drugs only if and when certain stimuli are present, using the decision-making capabilities of cells to supplant traditional pills with more sophisticated agents.

Returning to physical perspectives on the microbiome: other labs have also been uncovering biophysical drivers of gut microbial population dynamics. The lab of Terence Hwa in the physics department at the University of California at San Diego, for example, has built artificial devices that mimic aspects of the pulsatile flow of the human gut, recapitulating the natural ratios of abundances of characteristic bacterial species. Hyun Jung Kim at the University of Texas at Austin, makes stretchable gut-on-a-chip devices, similar to the lung-on-a-chip platforms of chapter 8, with which to study mechanical couplings to cultured intestinal cells. Other groups have looked at the chaining together of bacteria by antibodies, the mechanical roles of fiber and osmotic stresses, and more. I should stress that there are far more studies exploring the biochemical and genetic properties of gut microbes rather than their physical attributes, and those sorts of investigations are particularly well suited to discovering the means by which microbes communicate with each other and their host. Bacteria can synthesize unusual proteins, fats, hormones, vitamins, and even neurotransmitters. Undoubtedly, biological, chemical, and physical principles are all simultaneously at work in the gut microbiome.

THE SELF-ASSEMBLED ECOSYSTEM

We can think of the dynamics that emerge from the interplay of bacterial architecture and gut mechanics as another manifestation of our theme of self-assembly. There are even more unusual and abstract ways in which self-assembly emerges in microbial communities related to general properties of ecosystems. I mentioned earlier that your gut is home to many different species of microbes—hundreds, in fact. You'll also find an abundance of species in a bucket of seawater or a spoonful of soil. This diversity is puzzling. In fact, it's a classic conundrum in ecology, named "the paradox of the plankton" by G. Evelyn Hutchinson in 1961. The problem is this: Imagine some bunch of species and just one type of food available. There will always be one species that's better than the others at consuming the food and reproducing, becoming more abundant than the others by a margin that can only grow with time.

Eventually, we'll find an environment that's completely dominated by that species; rather than diversity, a monoculture. With a few different types of food present we can have a few coexisting species, but unless there's an amazing variety of food with an exquisite matching to particular species' preferences, highly diverse ecosystems should be impossible.

Nature, however, scoffs at this argument, and despite what theory dictates, routinely generates and maintains cacophonous diversity. Even in theory, however, there are many resolutions to the paradox of the plankton. One is spatial structure; if different creatures dwell in distinct zones, they can coexist even if their nutrient preferences are the same. Another is temporal structure; populations can oscillate out of sync, for example, so that one is dominant at one time and another at another time. Other resolutions relate to metabolism. In our simple picture above, an organism simply takes in food and reproduces. In reality, there are many intermediate steps. Molecules are converted into other molecules, broken apart, and joined together as they are consumed. Especially for bacteria, some of the intermediate molecules are secreted into or taken up from the surroundings. There are many more types of molecular nutrients available, therefore, than what we'd guess just by looking at the starting or ending points. Microbes feed each other through this chemical cross talk. Because of this, broth containing just a single type of nutrient can reliably sustain dozens of bacterial species. A few years ago, Alvaro Sanchez and colleagues at Yale University put together hundreds of microbial communities derived from samples of soil and plant leaves, feeding each on just one nutrient, and demonstrated not only that several species can routinely coexist but that the compositions of the resulting ensembles are predictable and reproducible.

Our theoretical understanding of ecological diversity is also advancing, offering deep insights into why coexistence can be more common than previously thought. One can write down mathematical equations describing the growth, death, and interactions of species in an ecosystem. Rather than considering particular values of the parameters in these models, one can assess the range of outcomes for many instances of randomly chosen parameters to get a sense of what sorts of properties are likely or unlikely to emerge. Again, our theme of predictable randomness shows up; rather than assessing the average properties of random walkers, we assess the average properties of random ecosystem models, evaluating, for example, how often they give outcomes in which all species are present or some number are extinct. This approach was pioneered by ecologist Robert May, who in classic and highly influential work in the 1970s concluded that the diversity and stability of ecosystems do not go hand in hand. Rather, addition of species to an ecosystem makes coexistence of the species less likely. Many theorists have continued along these paths, for example, Pankaj Mehta and colleagues at Boston University, who showed that coexistence can emerge beyond the destabilizing point theorized by May, though among particular subsets of the interacting species rather than all of them together.

Other theoretical approaches more explicitly relate nutrient use to the rise and fall of populations. Often called *consumer resource models*, these date to classic studies from the 1960s and 1970s by Robert MacArthur and other ecologists and gave rise, for example, to the paradox of the plankton noted above. Resolutions to the paradox, we now realize, can come from many sources. Simply adding constraints to nutrient use, even without cross-feeding, can be sufficient for species with similar nutrient preferences to coexist. The constraints aren't complex; imagine you can eat potatoes, carrots, and peas, but the total amounts are limited by the space on your plate, so any increase in the potato portion must be offset by a corresponding decrease in the area occupied by carrots and peas. Mapping onto metabolism, the vegetables represent different digestive enzymes, tailored to different nutrients, and the plate is the overall rate at which an organism can produce enzymes. Ned Wingreen and colleagues at Princeton University found that a mathematical description of this sort of constrained resource usage finds a surprisingly large set of parameters for which coexistence occurs, essentially because there are many ways to gently consume multiple foods that are equivalent in overall intake to ravenously consuming one food.

The field of theoretical ecology is vast, and the few paragraphs above certainly don't span its breadth. I picked this handful of examples in part because they illustrate important recent insights, but also because they illustrate yet again cross talk between physics and biology. The mathematics of Robert May's methods was that of *random matrix theory*, developed by physicist Eugene Wigner in the 1950s to make sense of the energy levels of heavy atoms by considering random parameter sets for the governing quantum mechanical equations rather than intractable exact solutions; May realized that the formalism could be translated to ecological systems. Pankaj Mehta and Ned Wingreen are both physicists, and their work draws upon theories of phase transitions and other physical systems to illuminate the mysteries of ecology.

Returning to my own work, my lab's observations in zebrafish have convinced me that physical structure and mechanical forces play major roles in the dynamics of the gut microbiome. There's no way we could have predicted the outcomes of *Vibrio cholerae* invasion, or antibiotic responses, by making measurements in a petri dish. The physical environment of the gut is as central to this ecosystem as the rocks and tides are to the coast. Of course, it's possible that everything we see is an idiosyncrasy of zebrafish and that our experiments don't teach us anything about humans or other animals. I think this is unlikely, though, not only because of the parsimony of nature that we've commented on before but also because the core underlying mechanisms are widespread throughout the animal kingdom. Every gut pushes material by mechanical contractions. Aggregation, in a gut or elsewhere, is one of the most common bacterial behaviors. It is difficult to imagine, therefore, that these dynamics would somehow disappear when applied to your own intestinal community.

Of course, the size of the gut, the numbers of species present, and the magnitudes of flows differ greatly between a larval fish and you. How do these attributes depend on the size of an animal? Can we think of your gastrointestinal tract as a scaled-up version of a fish's or a scaled-down version of an elephant's? We don't yet know. Despite enormous interest in the human gut microbiome and a sizable body of work on a few model animal species, there has been much less investigation of the gut microbiota of different animals. These questions, however, lead us to the even more general question of what "scaled" actually means. It turns out that there are general rules governing the variation of physical forces with size and shape, with dramatic consequences for differently sized organisms, and it is to these rules that we next direct our focus.

A Sense of Scale

Living creatures span a staggering range of sizes. A blue whale, a few tens of meters from head to tail, is about 10,000 times longer than an ant. Since van Leeuwenhoek in the seventeenth century, we've known that the ant and animals like it are not the extreme but rather the midpoint of nature's scale. Another factor of 10,000 separates the ant from the smallest bacterium. No less impressive is the variety of shapes that accompany this variety of sizes. Large organisms don't simply look like blown-up versions of small ones.

We wouldn't mistake the spindly legs of a rhinoceros beetle with the stumpy limbs of a rhinoceros, even if the former were stretched to the size of the latter. Photosynthetic algae are bulbous and compact; none bother with the riotous branching that trees are so fond of, despite their shared aim of capturing sunlight and carbon dioxide. These differences extend to behaviors as well as shapes: the back-and-forth stroke of a tail fin is typical of sharks, but we never see it while watching bacteria swim. We'll see why this is shortly.

As in earlier chapters, we can ask whether the amazing diversity of life coexists with underlying rules, and again the answer is yes. The size, shape, and behavior of animals are all intertwined, with relationships governed by the physical forces they encounter and the

156 • Chapter 10

environments they inhabit. A powerful concept that guides our understanding is *scaling*. Forces, flows of energy and matter, material characteristics, and geometric properties all have particular dependencies on size—they *scale* with size in particular ways—and these relationships dictate the forms and actions that life can harness.

HOW BIG IS A HORSE?

To tie size, geometry, and other characteristics to the ways animals and plants work, we're going to make use of a few mathematical tricks. One is a very liberating sort of inexactness.

How big is a horse? You already know the answer to this. We don't have to look up measurements of horse heights or books on equine anatomy. We needn't worry about whether I mean the distance from hoof to shoulder, or head to tail, or something else. We all know that a horse is about 1 meter in size. It's bigger than 0.1 meters (about 3 inches) and smaller than 10 meters (about 30 feet), whatever horse we picture in our minds. Whether our hypothetical horse is 1.0 or 1.5 or 2.53 meters tall matters if we're making a sweater for it, but not if we want to understand why a horse's bones are thicker and its metabolism slower than a 0.1 meter mouse's. Life spans vast extremes of sizes, and the relationships between size and function aren't determined by minor details.

Let's think of several different living things and note their sizes. To start, an ant is about a millimeter long, or 0.001 m; a typical virus is about 0.0000001 m in diameter.

It's tedious to write all the zeros in these numbers, and it's hard to keep track of how many zeros there are supposed to be. We therefore make use of scientific notation, denoting the powers of 10 in the number we're considering. Consider the number 100, which equals 10×10, in other words, 10^2. The number 10,000 is $10 \times 10 \times 10 \times 10$, or 10^4. Similarly, $1,000,000 = 10^6$ and $10 = 10^1$. What is 10 to the zero power, or 10^0? It's one fewer factor of 10 than 10^1, and so is 10^1 divided by 10, which is 1. Therefore, $10^0 = 1$. (By similar logic, *any* nonzero number raised to the zero power is 1.) What is 10^{-1}? We divide by another factor of 10, so $10^{-1} = 10^0/10 = 1/10$, which is 0.1. Similarly, $10^{-2} = 0.01$, $10^{-6} = 0.000001$, and so on.

You've almost certainly encountered scientific notation before. I describe it here for completeness but also to highlight the patterns by which we can construct relationships among the numbers. In the classes I teach for non-science-major college students, I often ask, "What is 10^0?" and nearly everyone answers "1." Few, however, can explain why. I ask them to imagine telling a friend that $10^0 = 1$, having the friend reply, "I don't believe you!," and then trying to persuade them of this mathematical relationship. Simply stating "That's the rule" isn't (or shouldn't be) a convincing argument, but describing patterns among numbers is. Better still, by understanding the patterns, you can construct the rules for yourself, whenever needed, without relying on rote memorization. It's liberating.

Returning to our list of living things, here's mine, with "order of magnitude" (i.e., powers of ten) values for characteristic sizes:

Object	Size
a molecule of DNA (width)	10^{-9} m
an antibody	10^{-8} m
an influenza (flu) virus	10^{-7} m
an *E. coli* bacterium	10^{-6} m
a red blood cell	10^{-5} m
a human egg cell	10^{-4} m
an ant	10^{-3} m
a blueberry	10^{-2} m
a rat	0.1 m
a horse, or a human	1 m
a blue whale (length)	10 m
a redwood tree (height)	10^2 m

You can make your own list, spanning its own wide range of powers of 10. How do the physical forces that animals and plants deal with vary, as we look up and down the ladder of sizes? Let's start by considering swimming.

WHY CAN'T A BACTERIUM SWIM LIKE A WHALE?

With elegant up-and-down strokes of its tail, a whale glides through the ocean. Sharks and many other fish similarly make use of a back-and-forth motion, in those cases left to right rather than up and down since their tail fins are vertical, but nonetheless *reciprocal*, meaning that the stroke in one direction retraces the path of the opposite stroke. If you peer through a microscope at a swimming bacterium, paramecium, or other microorganism, you never see that sort of swimming. You'll find a dazzling variety of motions—corkscrewing rotations of flagella, bulges that propagate along the cellular body, and more—but never a back-and-forth, reciprocal motion. Let's see why this is.

Every creature that swims through water pushes fluid as it moves. There are two reasons that pushing fluid is difficult. One is inertia: just as kicking a stationary soccer ball requires a force to accelerate it, it takes force to take an otherwise motionless bit of water and increase its speed. The second is viscosity: the honey pushed by a spoon drags against the honey alongside it, and it takes force to overcome this resistance. The ratio of these two required forces, the inertial force and the viscous force, is called the Reynolds number, after the fluid dynamics pioneer Osborne Reynolds, who became in 1868 the second ever "professor of engineering" in England. Every situation involving fluids has a Reynolds number, and the Reynolds number gives a concise way of characterizing the flow. High Reynolds number flows are turbulent—the inertia-dominated flows eddy and swirl, resulting in all the bits of water colliding chaotically with one another like little soccer balls. In contrast, low Reynolds number flows are smooth—the viscosity-dominated flows gently decay around the moving object. ("High" and "low" numbers are in comparison to 1, the Reynolds number at which the inertial and viscous forces are equal in magnitude.) We can calculate the Reynolds number given the properties of the moving object and the fluid it's in. High speed, large size, and low viscosity give a large Reynolds number; low speed, small size, and high viscosity give a low Reynolds number.

For a bacterium in water, with a size of about 10^{-6} m and a speed of about 10^{-5} meters per second, the associated Reynolds number is about 10^{-5}, or 0.00001—very, very low. For a cruising whale, the Reynolds number is about 10^8, very large, and 10,000,000,000,000 times greater than that of the bacterium. (You can see why we care only about orders of magnitude; worrying about whether the bacterium is 1×10^{-6} m long or 2.61×10^{-6} m long would be utterly irrelevant given the 13 powers-of-10 difference in the Reynolds number that we care about.) The bacterium and the whale, therefore, live in very different fluid worlds: the bacterium's is smooth and the whale's is turbulent.

In a classic 1977 paper, "Life at Low Reynolds Number," physicist Edward Purcell explained that this fact has surprisingly deep consequences for how aquatic creatures can and cannot move. At high Reynolds numbers, flows are irreversible, meaning that if we move an object one way through a fluid and then retrace its path to bring it back to its starting point, the fluid doesn't return to its original configuration. In other words, if you pour cream in your coffee, move your spoon through the liquid mixing together the coffee and cream, and then move your spoon back to its starting point tracing back the same route, the cream doesn't unmix. Being large, fast animals, we're very familiar with the high Reynolds number world; this irreversible behavior is so commonplace, we hardly think about it. (The spoon and coffee, by the way, have a Reynolds number around 1000; their motions drive the molecules of water and the oils and particulates of cream and coffee in chaotic, turbulent swirls.)

At low Reynolds numbers, flows are reversible. If I took the same cup of coffee and magically increased its viscosity by a factor of a million, its behavior would be dominated by viscous forces and hence would be reversible. Sweeping the spoon one way might seem to mix the cream and coffee, but by retracing the opposite path, each bit of liquid would also retrace its path, unmixing as the spoon returns to its starting point. When finished, we'd see a compact blob of cream in the coffee, just like it looked immediately after we poured it. Demonstrating this is one of my favorite in-class activities, which I perform with a ro-tating cylinder of very viscous corn syrup and dye, mixing and de-mixing as if by magic. (A classic video of this effect by fluid dynamicist G. I. Taylor is available online; see the references for a link.) What does this have to do with bacteria? Increasing viscosity gives us a low Reynolds number, but so does reducing size or reducing speed; a bacterium in water, as we noted, lives in a world of very low Reynolds numbers.

Purcell realized that because of the reversibility of low Reynolds number flow, microorganisms simply cannot swim using back-and-forth motions. It's not that they haven't found the genes to make it possible or haven't developed the right biochemical reactions, but rather that if you're small, the laws of physics make it impossible to get anywhere with reciprocal motion. If a bacterium waves some sort of stiff append-ages one way and thereby moves forward . . .

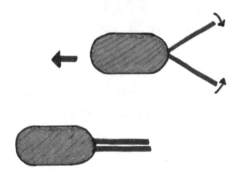

. . . it will move backward the exact same distance when it moves the appendages back:

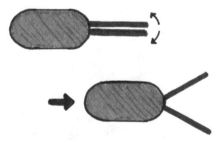

This is true even if the speeds of the back-and-forth motions are different, as long as the paths taken by the appendages are the same. Purcell named this the *scallop theorem* after the mollusks that move by clapping the two halves of their shell open and closed, a motion that would be futile if they were microorganisms.

What, then, do microorganisms do to swim? Anything they want, as long as it doesn't involve reciprocal motion. A common tactic is rotation of one or more helical flagella:

As long as the motor doesn't reverse, the movement of the flagellum never retraces its path backward and the creature steadily swims. Some microorganisms propagate bulges or kinks along their body, again making sure that the distortions don't retrace themselves. Another strategy is to wave hairlike cilia attached to a surface, making sure that the backward stroke isn't the opposite of the forward stroke. The cilia sway as they move one way . . .

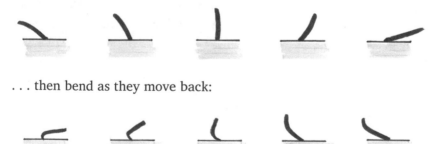

. . . then bend as they move back:

Though you're not microscopic, you make use of this motion a lot: cilia line the airways of your respiratory system, their motions propelling microbe-trapping mucus and transporting it away.

The world seen by the whale, therefore, is fundamentally different from that seen by the bacterium. Crossing the orders of magnitude in size that separate the two requires much more than shrinking or growing, but also changing the very form of the creatures' behavior.

SHAPES AND SCALING

As we've seen, differences in size can influence how animals must function. So too can differences in shape, and the two concerns are linked. We can develop insights into the complex shapes of animals by first considering some seemingly simple aspects of geometry. Suppose we have a square and we double the length of its sides. The area of the square has increased by a factor of 4, which we can see by drawing . . .

. . . or from the math that the original square's area is $L \times L = L^2$, where L is the length of a side, and the new square's area is $(2L)^2$, or $4L^2$. If we have a triangle, each of whose sides has the same length, and we double this length, the area again increases by a factor of 4; tripling the length of a side, the area increases by a factor of 3^2, in other words, 3×3, or 9.

This also holds for triangles with unequal sides, provided we stretch each side by the same factor.

In all the above cases, the area is proportional to the square of the length. Another way to state this is that the area scales as L^2, which we can write symbolically as $A \propto L^2$. Increasing the lengths by a factor of

2 means that we increase the area by a factor of 2^2, or 4. Replacing L with $3L$ means that we increase the area by a factor of $3^2=9$. Turning L into $4L$ increases the area by a factor of $4 \times 4 = 16$.

This probably seems quite basic. After all, you might say, we learned about the areas of simple shapes in elementary school. There is a subtle lesson here, however, that is not often stated. We don't need to know mathematical formulas for the areas of shapes. *Any* shape that doubles its lengths while keeping its form unchanged will quadruple its area. That's what area *is*; it's the geometric property that scales as L^2. A circle whose radius increases by a factor of 5 has an area that is $5^2=25$ times larger than it originally was, and there's no need to invoke the equation for the area of a circle. A sphere whose radius increases by a factor of 10 has a surface area that has grown by a factor of 100. The blob on the left has 4 times less area than the blob on the right, whose lateral extent is 2 times larger:

The *volume* of an object scales as length cubed, in other words, $L \times L \times L$, or L^3. You can convince yourself of this by drawing boxes (or other shapes, if you're ambitious) and showing that doubling all the lengths increases the volume by a factor of $2^3=8$, tripling the lengths increases the volume by $3^3=27$, and so on. Again, this is independent of what the shape is. If we multiply the radius of a sphere by a factor of 4, its volume increases by a factor of $4^3=4 \times 4 \times 4 = 64$. Halving the lengths of a three-dimensional blob, but keeping the blob shape the same, would give a new blob with one-eighth the volume.

Finally, we note that if we stretch or shrink an object, not altering its shape, the proportionality between any of the lengths doesn't change. Scaling up a triangle in a way that doubles its height also doubles its width. All lengths scale as L, which seems odd to write but which is good to keep in mind. Similarly, all areas are proportional to other areas; if we expand a shape so that its cross-sectional area increases by a factor of 4.7, its surface area also increases by a factor of 4.7.

The generality of all these aspects of scaling, whether of length, area, or volume, mean that we can apply them to questions of size and form of even the most complex organic shapes, as we'll see shortly. First, let's briefly revisit bacteria. We learned in chapter 9 that there are at least as many bacterial cells in your body as human cells—perhaps disturbing from the perspective of a census, but less so in terms of space. A typical bacterium is about 10 times smaller in width than a typical human cell. Its volume, therefore, is about $10^3 = 1000$ times smaller. Despite their large numbers, the microbes are dwarfed in volume by the human cells in your body.

Are the shapes of big animals similar to small ones? We can assess this in a more rigorous and quantifiable way than just visual observation. As we've seen, if shapes are similar, then their volumes scale as length cubed and their areas scale as length squared. We can flip this statement around: if a collection of animals shows volumes that are proportional to length cubed, or areas proportional to length squared, it tells us that in a general sense their shapes are similar. The technical term is that they show *isometric scaling.* If the animals' volumes, for example, aren't proportional to the cubes of their heights, perhaps becoming disproportionately chunky when large or not showing any consistent relationship at all, we know that nature has discarded isometry, hinting that other concerns are at work. The challenge, then, is to assess the scaling behavior of real-life animal forms. One can do this with equations, but the easiest and most insightful approach is visual, with a tool that forms our second mathematical trick: logarithmic graphs.

Suppose we make a graph of volume versus length of a side for cubes. The usual way to plot this would look like the graph on the left, swooping upward with a cubic form.

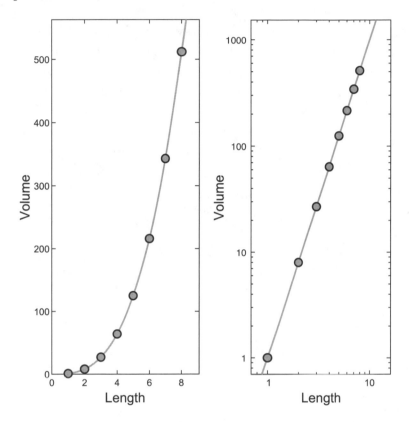

If we instead plot the same numbers on a different sort of graph paper, in which the equally spaced divisions are powers of 10 (right graph), we find something remarkable: the points lie on a straight line. What's more, the slope of the line—the number of "vertical" powers of 10 divided by the number of "horizontal" powers of 10—is 3. If we had plotted surface area rather than volume, we'd again find a straight line, but with a slope of 2. Quite generally, if y is proportional to x^p, where p is some exponent, plotting y versus x on logarithmic axes gives us a straight line of slope p. We can simply read off the scaling exponent from the graph. This ability to discern scaling relationships visually, from the tilt of the trend on an appropriately constructed plot, delivers all sorts of insights into animal shape and form.

A COCKROACH IS A COCKROACH IS A COCKROACH

Returning to animals and armed with our graphical tools, we can ask the burning question, Are cockroaches isometric? It may seem a silly thing to ask, but the answer can reveal what mechanisms of development shape the growing animal, how mechanical stresses affect different members of a species, and even how new species evolve. Rather than plotting volume, researchers often consider the mass of an animal, which is easy to measure by weighing. Most animals have similar densities of their constituent cells and tissues, so mass is roughly proportional to volume. I've reproduced here a plot of mass versus leg length for several different cockroaches, big and small.

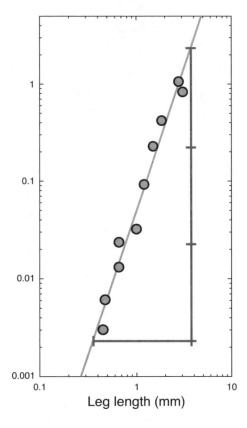

The data aren't mine—as fond as I am of nature, I can't stand cockroaches—but come from a 1977 paper by biologist Henry Prange, examining the mechanics of exoskeletons. The points are exceptionally well fit by a straight line. Of course, animals are not as well behaved as cubes or spheres, and the data are scattered about a line rather than lying perfectly on it. Still, the best-fit slope is 2.95, almost exactly 3, indicating that mass scales as length cubed just as we'd expect for isometric objects. (I've included on the graph horizontal and vertical lines with equally spaced tick marks, making it more evident that the slope is 3.) The large cockroaches are just like scaled-up versions of the small cockroaches; the size changes, but the shape is essentially the same. The cockroaches, in other words, are isometric.

Here's another example of isometry, this time across many different animals. All mammals have lungs with which to bring into their bodies fresh, oxygen-rich air. We can ask how the size of the lungs depends on the overall size of the animal. By "size," we could mean surface area or we could mean volume. In the next chapter, we consider surface area, by far the more interesting of the two measures, and how it's deeply connected to the ways lungs work or sometimes fail to work. Here we consider the more boring measure, volume, as a prelude.

We might expect isometry from mammalian lung volumes if, for example, every cell needs a similar volume of oxygen at each breath, so that the total volume of air is proportional to the total volume of cells in the body; in other words, lung volume scales with animal volume. Or we might guess that large animals need disproportionately more or less oxygen volume per cell, in which case the volume of the lungs wouldn't be proportional to the volume of the body. In the first case, plotting one volume versus the other on a logarithmic graph gives a slope of 1; in the second, it doesn't. Let's make the plot, taking values from a 1963 paper on lung physiology. As before, we plot the mass of each animal rather than its volume. Across mice, monkeys, manatees, and more, we find a slope of 1 (1.02 to be exact), indicating isometric scaling of the lungs. As mammals get bigger, their lung volume gets bigger by the same proportion. It's tempting to conclude that the typical cell in each animal is satisfied by roughly the same volume of air, as suggested above. It could be, however, that lung volume isn't the only factor that sets the availability of oxygen, and in fact we'll see in

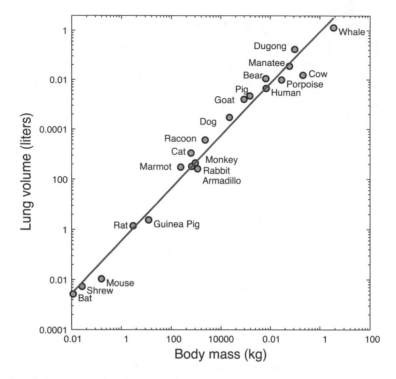

each of the next two chapters nonisometric scaling governing oxygen transport and metabolism. It's worth keeping in mind that logarithmic graphs and scaling exponents don't in themselves provide explanations, let alone correct ones. They can, however, point us toward insights that might not otherwise be apparent, as the next example shows.

WHY ELEPHANTS CAN'T JUMP

As we've seen, nature sometimes obeys isometry. It doesn't always do so, however! The elephant and the elephant shrew look very dissimilar, despite sharing a long nose. Many measures fail the test of isometry, sometimes revealing other rules. To illustrate, let's consider the large family of cloven-hoofed grazing mammals known as bovids, which includes massive water buffalo, midsized goats and sheep, and antelopes ranging down to the foot-tall (30 cm) dik-dik. In their excellent book *On Size and Life*, Thomas McMahon and John Tyler Bonner plot on logarithmic axes the diameter versus the end-to-end length of the humerus, a leg bone, of lots of bovids. I've reproduced the graph here. If these animal skeletons were isometric, these data points would have a best-fit line with a slope of 1. (Diameter and end-to-end length are both measures of length and should therefore be proportional to each other for shapes that stay the same.) The graph instead has two striking features.

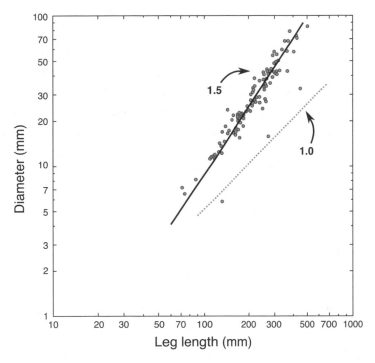

First, there *is* a relationship between diameter and length—the points aren't just randomly arrayed across the page. Second, the relationship is *not* isometric. The best-fit line has a slope of 1.5, unmistakably different from 1.0. (I've drawn a dotted line of slope 1 for comparison.) The bovids, therefore, have abandoned isometry, with the larger bovids sporting disproportionately thick bones compared to their smaller cousins. If we double the length of a bovid bone, we must more than double its diameter; in fact, the diameter must increase by a factor of $2^{1.5}$, which is about 2.8. The bovid skeletons are not similarly shaped versions of each other but have some other rule governing their form.

What is this rule, and why an exponent of 1.5? The physics behind it has to do with gravity, tugging animals and everything else downward, and the strength of bones, holding the animals up. The force of gravity is proportional to the mass of an object. If the mass increases by a factor of 10, the force of gravity increases by a factor of 10. The strength of a bone, specifically the maximum force it can withstand, depends on its cross-sectional area. This isn't a peculiarity of bones but rather a general statement about the mechanics of any sort of beam. Given these laws, our bovids would face a terrible problem if they were isometric. If we doubled, for example, all the lengths of an antelope's anatomy while keeping its shape the same, its volume would increase by a factor of 2^3, or 8. The animal's mass, and therefore the force of gravity acting on it, would also increase by a factor of 8. However, the cross-sectional area of the isometric animal's bones, like any other area, would only increase by a factor of 2^2, or 4. The strength of the bones increases less than the weight it must support. There are two possible responses to this. One is for the big animals to give up on having bone strength (relative to their weight) similar to the small animals, and to adopt correspondingly different lifestyles. The other is to have disproportionately thick bones, abandoning the similar shapes of isometry in favor of bones that are wide enough for their strength to balance their gravitational demands. The bovids take the latter path. Rather than similarity of shape, McMahon and Bonner and others posit that *elastic similarity* governs the bovid bones—the amount the bones may bend, relative to gravitational stresses, is roughly constant across different species. It turns out that a diameter-to-length scaling exponent of 1.5 corresponds to elastic similarity; if the bones become extrathick in just this way, their resistance to bending is like that of the smaller bones, relative to the force that's bending them. In other words, if the small bovids can walk, run, and graze, the large ones can too, in similar ways, thanks to their skeletal architectures scaling according to elastic similarity rather than isometry. (There's a lot of complexity that I'm glossing over, as well as a justification that bending rather than other failure modes is the critical factor for skeletons.) Of course, the bovids don't make logarithmic graphs or perform mechanical calculations. Rather, the evolutionary demands of being a bovid selected for animals whose bones obeyed the geometric relationship illustrated above. Those with smaller bone diameters were too weak to withstand gravity; those with diameters that scaled even more sharply than length to the 1.5 power perhaps expended much more energy on growing chunky bones than was necessary for a bovid lifestyle. The bovids, without knowing it, took a path constrained by biology and physics to reach an elegant endpoint of nonisometric scaling.

I mentioned, however, that there's another path that large animals can take as they consider the balance between bone strength and gravity: giving up on having the same behaviors as small animals. This is, in fact, what large animals in general do. Plotting on logarithmic axes the diameter versus the length of leg bones for a wide array of animals—cats, dogs, horses, elephants, and more—gives a scatter of points with a crude slope that's steeper than the isometric value of 1.0, but not as steep as 1.5. The bones of an elephant are truly impressive: my students and I have gathered in awe around the femur of an elephant named Tusko that I've carted into class. It's massive: a meter long with the girth of a small palm tree. (Tusko, by the way, lived a mostly miserable life in early twentieth-century circuses until ending up under far more pleasant care at the Woodland Park Zoo in Seattle. After his death, his skeleton was donated to the University of Oregon, where the elephant lives on, teaching students about bones.) Alongside Tusko's leg bone I show a coyote femur, far smaller at just a centimeter wide and 12 centimeters long. The elephant's femur is 9 times longer and 16 times wider than the coyote's. Tusko's bone is thicker than isometry would dictate (which would be just 9 times wider), but not as thick as elastic similarity would demand (a factor of $9^{1.5}$, which is 27, times wider). We can't fault the elephants for not maintaining elastic similarity with the smaller animals. If they did, bones would make up about three-quarters of their body mass, which would likely cause all sorts of anatomical complications. The consequence, however, is that the bones of the large animals are not as strong, relative to the animals' weight, as those of the small animals. The coyote can jump; the elephant cannot. The extra stress would shatter its bones. It's for this reason that elephant enclosures in zoos are often surrounded by simple, narrow moats that many animals—but not the elephant—could easily leap across.

The observation that bigger animals have disproportionately thick bones, and the assessment that they're still not thick enough to match the relative strength of small animals, are both quite old. Galileo, in fact, wrote about bone scaling in his 1638 book, *Two New Sciences*. One of the characters who animates the book speculates, "I believe that a little dog might carry on his back two or three dogs of the same size, whereas I doubt if a horse could carry even one horse of his own size." Though Galileo is renowned for performing insightful experiments, I don't think he actually tried this one.

The realization that animals' sizes, forms, and activities are all inter-related, and that these relationships are often set by the physical forces manifest in the environment, gives us deep insights into the natural world. There's a unity to all the bovids in their elastic similarity, for example, that's even more profound than their superficial similarities would imply. The back-and-forth swaying of a shark's tail is a testament to its liberation from the fluid constraints that apply to smaller crea-tures. The repeated emergence of scaling relationships as we study the living world across its wide range of sizes elevates the notion of scaling to a broad guiding theme. Scaling concepts provide a coherence that unifies the diversity of animal shapes and behaviors. Scaling also helps explain why such diversity exists, as organisms evolve amid physical forces whose manifestations and magnitudes are different at different sizes. In the next chapter, we focus on scaling issues related to surfaces, especially the challenges of breathing.

Life at the Surface

You have lungs. Ants don't. Why?

The ants' cells, like yours, need oxygen. Rather than lungs or any analogous organ, they make do with a few tiny holes along the sides of their bodies, one pair per body segment, connected to tubes on the inside. It's not because of some marvels of cellular engineering that small insects can bypass the need for lungs. Instead, it's because of scaling, and the physics and geometry of surfaces. We explore in this chapter the interfaces between inside and outside, or between one space and another, universal features of which govern the shape of an elephant's ears, the challenges of a baby's first breath, and more.

ABANDONING ISOMETRY

Every creature exchanges oxygen and carbon dioxide at some surface. The rate at which it can do so is limited by the surface area available. There's nothing complicated behind this—gas molecules going from one environment, such as the air, to another, such as the interior of a blood vessel, have to cross the boundary between them, so the flow of molecules is proportional to the surface area of that boundary. The overall demand for oxygen is set by the creature's volume (deferring some complications we consider in the next chapter), as each of its cells consumes oxygen to perform the chemistry of respiration. For a small creature, such as an ant, the surface area of its internal tubing suffices for gas exchange with its tissues, and the animal's outer surface area suffices for capturing the oxygen that its volume needs.

Imagine, however, magically expanding an ant, isometrically. If we double all the lengths of an ant, making it 6 millimeters long rather than 3, its volume increases by a factor of eight, as we saw in the last chapter. The number of cells also increases by a factor of eight, and the total amount of oxygen it needs increases by a factor of eight. The surface area, however, increases by a smaller factor, namely, four. As we increase size further or imagine even larger animals, the discrepancy gets even greater. A human-sized ant, 1000 times longer, would demand 1000^3, or 1 billion (10^9), times as much oxygen, but would have only 1000^2, or 1 million (10^6), times as much surface area with which to supply it. At human size, one can't exchange enough oxygen and carbon dioxide at the surfaces of simple scaled-up tubes, nor can one absorb the requisite amount of air through a passive outer surface. What was easy to do when small becomes impossible when large, simply because of geometry.

All large animals and plants, you included, deal with this challenge by abandoning isometry. Instead of keeping shapes similar, larger organisms adopt forms that have much more surface area for gas transport, enough to keep up with the increase in volume and cellular mass.

Small photosynthetic bacteria are smooth and round. Trees, in contrast, are highly branched, with an abundance of leafy surfaces at which to exchange carbon dioxide and oxygen. You, too, have branches, but on the inside: you have lungs, the airways of which divide into smaller and smaller airways in a cascade of ever-finer features, resulting in an enormous internal surface area. It's often noted that your lungs have the volume of a few tennis balls and the area of a tennis court.

The actions of your organs are also rooted in scaling. Your lungs inflate and deflate, driving gas in and out of your body. The ant, in contrast, can simply rely on the holes at its surface. Though ants and other larger insects can expand and contract parts of their bodies to actively pump air, the dominant mode of gas transport—and the only mode present in smaller insects—is passive diffusion. This won't suffice for larger creatures; as we saw in chapter 6, diffusing over long distances is very slow, and so you'd suffocate if you stopped pumping. Similarly, your heart vigorously drives oxygenated blood through another branching network spanning your body. The ant doesn't bother; internal gas exchange is also made simple by its small size. A little creature can harness the predictable randomness of Brownian motion, as air that enters its superficial holes meanders through passageways that penetrate the body. Invoking the insights of chapter 6: the random walk of a particle, such as a molecule of oxygen, travels on average a distance that grows as the square root of its travel time. Inverting this relationship and using the language of scaling, the travel time scales as length squared. Being 1000 times larger than the ant demands 1000^2 more time for diffusion, in other words, an extra factor of one million. Rather than patiently enduring this millionfold slowness to oxygenate our tissues, large creatures like us carry oxygen in blood and force the blood through a circulatory system, bringing it close enough to each cell for diffusion to quickly carry it the rest of the way.

There is, by the way, a way to be big and to have sufficient oxygen without building respiratory organs with large surface areas, and that is to live in an environment with a highly oxygen-rich atmosphere. Such a place doesn't presently exist, but there were times when it was the norm. In the Carboniferous period about 300 million years ago, for example, the ambient oxygen concentration in the air was 50% higher than it is currently, and we find in the fossil record abundant evidence of giant insects. A prehistoric dragonfly with a 2-foot (60 cm) wingspan would have trouble with today's relatively impoverished air.

Surfaces influence many aspects of animal form. Moose are bigger in colder regions. Bears are larger as well; polar bears, closely related to brown bears, are more massive than their southern cousins. This general relationship that animals of a species tend to be larger at colder latitudes has been noticed for centuries. The likely explanation is surface area. If you're warm blooded and live in a cold place, your body's surface area is a liability—that's where you lose heat. Because surface area scales as length squared, and volume as length cubed, the ratio of surface area to volume decreases with size. If an animal's internal heat production is matched by the heat loss through its skin and we isometrically double the animal's size, it now produces eight times as much heat from its eight times greater mass, but its rate of heat loss has only gone up by a factor of four. It could therefore overheat or, more realistically, it would require and consume fewer calories to sustain its body temperature. The larger animal can therefore survive more easily, giving rise to an evolutionary advantage for increased size. All other things being equal, being big helps in cold places.

An animal in a hot climate, on the other hand, must worry about overheating, and it benefits from having more surface area at which to dissipate heat. The ratio of surface area to volume is greater for smaller size, so all other things being equal, being small helps in warm places. Of course, one could instead abandon isometry, as with the elephant's giant-surface-area ears; but within species, changes of form tend to be less dramatic—hence the general observation about sizes and latitudes, named *Bergmann's rule* after a nineteenth-century biologist.

So far, our examples of surface-related principles relate to the form of animals. Surfaces also affect behavior, and what creatures can and cannot do.

WALKING ON WATER

At the top of a tranquil pond, water striders and many other insects effortlessly dance on the liquid, as easily as you might stroll on a lawn. Why aren't you similarly able to walk on water? The magic of a water strider lies not in the makeup of its legs but rather in its size. The insect's abilities are a consequence of scaling, specifically the scaling associated with a force called *surface tension*.

Surface tension arises at any liquid surface. No matter the liquid, the molecules making it up attract one another. That's an inherent attribute of liquids—if molecules don't attract one another, they'd very likely form a gas. Every water molecule wants to be next to other water molecules. Every oil molecule wants to be next to other oil molecules. Any liquid molecules at a surface, such as the surface of a pond, have about half as many neighbors as do molecules in the bulk. To anthropomorphize for a moment, the surface molecules are unhappy, and the liquid as a whole minimizes its surface area to ensure as few unhappy molecules as possible. Furthermore, the liquid resists any process that increases its surface area; the resulting force is known as surface tension. For soap bubbles, or liquids floating in space, or droplets in an oil and water vinaigrette, surface tension pulls the fluid into spheres because a sphere is the three-dimensional shape with the smallest surface area for a given volume. Water in a bucket, or water in a pond, has gravity and the container walls as additional constraints, and a flat interface with the air minimizes the available surface area. Whatever the context, we can think of every liquid surface as constantly pulling, contracting itself to have the smallest surface area possible given the constraints of its volume and whatever else is acting on it.

Now we can understand why the water strider can stride on water. Its legs push the water's surface, driven downward by gravity tugging on the insect. The legs are hydrophobic; the water molecules don't have any particular affinity for them, preferring instead to be near each other, again minimizing the overall surface area as much as possible. The insect's spindly legs deform the water surface; the water responds with the force of surface tension, trying to push the interface back toward a flat shape. If we imagine the leg moving downward as it alights on the pond, pulled by the force of gravity, the deformation steadily increases, and the upward force of surface tension also increases.

One of two things then happens: If we reach a deformation at which the force of surface tension balances the force of gravity, the insect stays above the water, not breaking the liquid surface, held up by the affinity of the water for itself. Or, if the gravitational force outweighs the maximum force that surface tension can provide, the surface breaks and the insect is submerged. Thankfully for the water striders, their evolutionary history has led them to the first outcome. This fluid-based support is easy to demonstrate, by the way, with a metal paper clip gently laid on a water surface. As long as both are very clean, the clip will be supported by the fluid despite being much denser. If you push the paper clip under the water's surface, however, it will sink—surface tension only applies at the surface.

So far, this argument explains why the water strider can walk on water, but it isn't clear why it fails to apply to you. After all, even though the force of gravity acting on your body is far greater than that on the water strider, your contact with the liquid surface is also larger than that of the insect. Shouldn't the upward force be greater, too? It is, but it's not enough to keep you standing atop a swimming pool. The reason once again lies in scaling. The force of gravity, as we noted in the last chapter, is proportional to the mass of an object and therefore scales with volume, or length cubed. The force of surface tension scales not as length cubed, or even length squared, but just *length*. A cube 1 inch on a side, perched on water, has a 4-inch length of perimeter defining the contact zone.

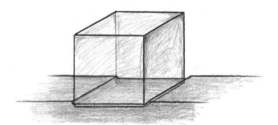

A 2-inch cube has 8 inches of perimeter, a factor of two larger than the original cube. It is along this perimeter that the water's surface is curved and extended compared to a flat interface, and so it is the scaling of the edge length that governs the scaling of surface tension forces. A creature 10 times as long as another, all other things being equal, has 1000 times the gravitational force pulling it downward, but can only muster 10 times the upward force from the fluid. A small creature may be fine at a liquid surface, but if we imagine it growing, it very quickly reaches the point at which gravity is far too strong to resist. This cross-over occurs at a size of a few millimeters. Below this, it's not hard for a creature to support itself by surface tension on water; above this size, it's hopeless.

Certain ants provide another illustration of the consequences of surface tension. Fire ants are an aggressive set of species with a painful, burning sting, hence the name. They are native to tropical areas prone to heavy rains that can flood their habitats, at which point they need to stay above the water and also stay together as a colony. Though ants are denser than water, an individual ant is small; like the water strider, it can stay atop water using surface tension. A group of ants clinging to one another to stay together, however, faces a problem. As the group grows, its mass increases more strongly than the surface tension force pushing it up via the same scaling that we've just seen. For more than a few dozen ants, gravity overwhelms surface tension and the group will sink. It helps only slightly to form a two-dimensional raft rather than a three-dimensional blob; the mass of the flat raft scales as length squared, which still quickly surpasses the length-to-the-first-power scaling of surface tension. The ants, it seems, are doomed by the physics of scaling to either disperse from one another or drown. They've devised, however, an ingenious solution to this dilemma: air bubbles. An ant's surface is hydrophobic, and an individual ant can clasp a bubble to its body like a mother might hold an infant. A raft of ants together hangs on to a large bubble of air; the buoyant bubble counteracts the gravitational force pulling the insects downward. Since surface tension is fundamentally incapable of doing the job, the ants make use of the low density of air to manipulate the gravitational side of the equation.

Other creatures manipulate surface tension more directly. You, in fact, are one of them. Our next example takes place within you, every time you breathe.

BREATHING IS HARD WORK

On August 7, 1963, Patrick Bouvier Kennedy was born, five and a half weeks early, to President John F. Kennedy and First Lady Jacqueline Kennedy. Within two short days spent struggling for breath, he died.

The tragedy was mourned by millions, and though the context of being a child of the president of the United States was special, the cause of the baby's death was frighteningly commonplace. Young Patrick died from infant respiratory distress syndrome, or IRDS, the leading cause of death among premature infants. IRDS is a problem of surfaces.

For every breath you inhale, you expend a lot of work to inflate your lungs. Lungs are often depicted as elastic sacs, like rubber balloons, stretched by muscles. Your lungs aren't just balloons, however; they're *wet* balloons. The cells that line each of the hundreds of millions of tiny air sacs that together make up your lungs coat themselves with a thin layer of liquid mucus (blue in the illustration). Inhalation stretches the rubbery tissue and also increases the area of the liquid surface, that is, the area of the interface between the inflating air and the fluid film. As always, the liquid "wants" to keep this area minimal and thus will oppose its growth.

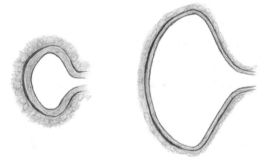

Both the stretching of the lung tissue and the expansion of its liquid surface take effort, and in fact the energy required is comparable for the two. One can measure this by pumping freshly dissected lungs with air or water. Filling with water, there's no air-water interface and hence no surface tension to fight, so the work required is solely due to stretching the lung tissue. Filling with air, we stretch the tissue and expand the air-water interface, which takes about twice as much work as filling with water. In other words, about half the effort of breathing is due to surface tension. This isn't too surprising—as we've noted, your lungs have an enormous surface area.

The high price of breathing would be even higher if the lung surfaces were lined with pure water. All liquids pull, but some pull more strongly than others. Among common liquids, water's surface tension is one of the largest—about twice as strong as oils or alcohols, a consequence of strongly attractive forces between its molecules. We can lower the surface tension of a water interface by adding a tiny bit of soap. You can demonstrate this with the paper clip that we placed on a water surface a few pages ago—with just a touch of dish soap, it sinks, its weight overwhelming the force that the soapy interface can muster. Soap accomplishes this feat by virtue of its molecular structure. As noted in chapter 5, each soap molecule has one end that's hydrophobic and one that's hydrophilic, just like lipids. The soap, therefore, gladly goes to the water's surface, where the hydrophobic tails jut out into the air. The surface is then no longer the domain of water molecules that would rather not be there. There's no longer such a large energetic cost for surface area, and surface tension is greatly reduced.

Returning to lungs, nature cleverly mitigates the cost of their expansion by simply adding soap. The secretions of the cells lining the lungs technically have the more impressive name of *pulmonary surfactants*, but "soap" is a better description of this mostly lipid substance. This is another manifestation of self-assembly: without any external guidance, the secreted soap organizes itself at the liquid-air interface, forming a layer of molecules that assists the entire organ.

The crucial ability to dramatically lower the surface tension of your lungs isn't one you've always possessed: pulmonary surfactant is made rather late in embryonic development. Premature infants, depending on how early they are, enter the world with a shortage of surface-modifying secretions or even with none at all; breathing may be a struggle or even impossible, as the muscles fight to defeat surface tension.

Patrick Kennedy wasn't the only child of his time to succumb to infant respiratory distress syndrome. In the United States alone, IRDS claimed 25,000 infants per year in the 1960s. By 2005, however, the annual mortality rate had plunged to less than 900. IRDS still occurs, just as before, but it is straightforward to treat: we squirt soap into the infant's lungs. Technically, of course, it's pulmonary surfactant, either extracted from animals or chemically synthesized, but this detail shouldn't distract us from the treatment's wonderful and effective simplicity, based not on complex biochemistry or genetics but rather on the physics of breathing. Self-assembly underlies this simplicity: the surfactant molecules put themselves in place, each positioned at the two-dimensional interface between liquid and air, on their own. Understanding surface tension saves lives.

Surfaces and the scaling associated with them govern many aspects of the living world and also illustrate how scaling connects the microscopic (for example, the structure of lipid molecules) and the macroscopic (for example, the mechanics of expanding your lungs). As in the previous chapter, we've seen how scaling helps us make sense of phenomena we observe in ourselves and other organisms. Scaling isn't a panacea, however, and we turn next to an example we're still struggling to make sense of.

12

Mysteries of Size and Shape

In the preceding chapters, we've seen examples of scaling relationships—connections between properties like bone strength and measures of overall size. These relationships often transcend simple proportionality; in surprisingly many cases, they are well described by dependencies of one measure on another raised to a power or exponent, a form generally known as a *power law*. Scaling illuminates many features of the living world. There are many examples beyond what we've seen so far, spanning sea, land, and air. Measuring the swimming speed and stroke frequency of a vast range of aquatic organisms, we find that the two are linked by an exponent whose value can be explained by hydrodynamics. Among animals that run, from cockroaches to horses, there is a power law scaling relationship between the energetic cost of locomotion and body mass, set by physical mechanisms deeper than the peculiarities of particular gaits. Flight speed, power, and mass are all governed by aerodynamic scaling laws that unify not only flying creatures but jet planes as well. We could explore all of these and more, but instead we'll take our successes for granted and look at something more puzzling. (If you're curious about these examples, I've listed some readings in the references.) We'll consider a scaling-related mystery that, if we could solve it, might help explain what sets the tempo of heartbeats, the biodiversity of forests, and even the limits of life spans.

ENERGY AND SIZE

Our mystery is old, controversial, and surprisingly easy to describe. Every organism extracts energy from chemical bonds. These bonds are present in the food consumed by animals, the nutrient molecules scooped up by bacteria, and the sugars photosynthetic creatures craft using sunlight. The released energy is directed toward all the tasks associated with life: growth, development, motion, reproduction, and more. The rate at which energy is used is known as the metabolic rate, the term *metabolism* referring to all the varied chemical reactions conducted by cells in the course of their activities. This rate isn't constant—when we sprint, we use energy much more rapidly than when we sleep—and it's never zero. Even at rest, every creature is using chemical energy at a rate called the *basal metabolic rate*. Different organisms have different basal metabolic rates. A resting elephant, for example, consumes more calories per minute than a resting mouse. How much more? More generally, how does the basal metabolic rate of an animal scale with its mass?

We might expect that no sensible relationship would hold across the diversity of life, with every creature's metabolic rate set by the uniqueness of its anatomy. Or we might expect that the basal metabolic rate would be proportional to body mass—an animal with twice the stuff inside it uses twice as much energy every minute.

In fact, neither of these is true. There *is* a relationship between basal metabolic rate and body mass, but it's not one of simple proportionality. For bone shape (chapter 10), we see a trend of disproportionately thick bones for larger animals. For basal metabolic rate, we find a trend of disproportionately lower energy use for larger animals. Basal metabolic rate is often expressed as the rate of oxygen consumption, because oxygen is required by the chemical reactions that drive metabolism. A resting mouse uses oxygen at a rate of about 40 milliliters per hour. An elephant, whose mass is 100,000 times greater, doesn't use 100,000 times as much oxygen as the mouse, but less than 10,000 times as much. A gram of resting elephant consumes energy at a rate about 20 times lower than a gram of resting mouse, on average. Lifting an evocative description from biochemist and writer Nick Lane, "an elephant-sized pile of mice would consume twenty times more food and oxygen every minute than the elephant does itself." This isn't a peculiarity of mice or elephants. In general, the larger the creature, the lower the metabolic rate per gram of body mass.

How much lower? In a 1932 paper that launched a thousand studies, the Swiss American physiologist Max Kleiber considered a range of animals from doves to cattle, plotting their metabolic rates and body masses on logarithmic axes. As we saw in chapter 10, this graphical form allows us to clearly see relationships described by exponents. A hypothetical basal metabolic rate that is proportional to mass (i.e., mass to the first power) would show up as a line with a slope of exactly 1. Kleiber didn't find a random scatter of points, nor did he find a slope of 1. Rather, the metabolic rate rose with body mass with a well-defined slope, and therefore a scaling exponent, of 0.75. Increasing an animal's mass by a factor of 100,000, for example, increases energy consumption by a factor of $100,000^{0.75}$, which is about 5600.

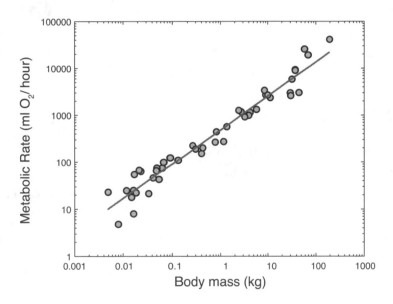

Many subsequent researchers added data points from more animals, expanding on Kleiber's set of 13. To illustrate, I've plotted about 50 points from a 2003 dataset of over 600 mammals, randomly chosen but roughly evenly spread across the wide range of masses, from the sooty mustached bat (5 grams, about the weight of a piece of paper) to the common wildebeest (200 kilograms). The linearity of the points is suggestive of some simple underlying law.

No one knows, however, what this law is. In the 90 years since Kleiber's observations, elevated to the status of "Kleiber's law" in countless textbooks and research papers, there has been no shortage of explanations, discussions, and arguments. These have not, however, converged to any agreement about what drives the relationship.

The simplest model that one might construct, beyond direct proportionality to mass, is that basal metabolic rate is proportional to the surface area of an animal. Echoing chapter 11, we might expect that the outflow of energy as heat depends on the external surface and that this must be balanced by the body's overall energy consumption rate. After all, nearly every activity of cells, tissues, and whole organisms ends up dissipating energy as heat, a consequence of universal laws of thermodynamics. This possible equivalence of metabolism and surface area predicts a basal metabolic rate that scales as mass to the ⅔, or 0.67, power. (If you're comfortable with algebra, this is because surface area scales as $length^2$, and mass as $length^3$. Inverting the latter relationship, length scales as $mass^{1/3}$, so surface area is proportional to $(mass^{1/3})^2$, or $mass^{2/3}$. If you're not comfortable with algebra, pretend you didn't read this.) This surface area model doesn't match the data, as we've seen; 0.67 is notably different from 0.75. A $mass^{2/3}$ scaling does, however, show up in several, though not all, studies of animals *within* the same species. In fact, 50 years before Kleiber, the German scientist Max Rubner examined the energy consumption of seven dog breeds and found a $mass^{2/3}$ scaling. Oxygen consumption in guinea pigs shows the same behavior. As in chapter 10, the data suggest that isometry is a good rough guide if considering individuals of the same species, but that the differentiation of species requires more drastic changes.

If surface area doesn't explain Kleiber's law, what does? One possibility is that Kleiber's law isn't really much of a law. Basal metabolic rate is a difficult quantity to measure, so we perhaps shouldn't place too much confidence in the data points. Furthermore, the points aren't arranged in a tidy row but are scattered quite widely about the best-fit line; we perhaps shouldn't trust its slope too much. In fact, the best-fit exponent for the 600 mammals noted earlier is 0.73, not exactly equal to the appealingly elegant ¾ (i.e., 0.75). Several expansions and reassessments of the data cast additional doubt on simple conclusions. In 2001, Peter Dodds, Dan Rothman, and Joshua Weitz at the Massachusetts Institute of Technology evaluated several datasets and concluded that values for small animals, less than about 10 kilograms (22 pounds, roughly the mass of a bobcat), were well fit by a scaling exponent of 0.67. Larger animals deviate upward from this trend. If one insists on fitting a single line to all the data points, an exponent around 0.71 emerges, which can be pushed closer to ¾ by evaluating a higher fraction of large mammals. Considering birds alone, Peter Bennett and Paul H. Harvey in the 1980s found a scaling exponent of 0.67, consistent with basal metabolic rate being determined by surface area. Studies of reptiles, however, give an exponent of 0.80.

Different sets of creatures yield different metabolic scaling relationships, calling into question the very notion that there is any sort of universal principle governing all of them. On the other hand, one might argue that these variations are to be expected, due to the natural peculiarities of organisms with different behaviors and life histories, and it is precisely the overall slope—the slope that emerges if we toss all of their data points onto one graph and squint to blur the discrepancies—that matters for insights into general principles. The latter point of view would be more compelling if we could state what those general principles might be.

For decades, debates about Kleiber's law simmered; but they were brought to a boil in 1997 by physicist Geoffrey West at Los Alamos National Laboratory and ecologists James Brown and Brian Enquist at the University of New Mexico, all three of whom shared an affiliation with the interdisciplinary Santa Fe Institute. West, Brown, and Enquist proposed a creative explanation for metabolic scaling, and moreover one that they claimed leads inexorably to a ¾ scaling exponent. In their picture, the key determinant of metabolism isn't the energetic demands of an organism's cells but rather the physical constraints of the circulatory and respiratory systems that supply them. As with the surface area argument and its associated ⅔ scaling, geometry is central to the theory. Unlike the familiar geometry of surface area, however, West, Brown, and Enquist's model made use of a different sort of mathematics known as fractal geometry.

FRACTAL BIOLOGY

There is an almost rhythmic complexity to the shape of cracks in a sidewalk, veins on a leaf, or frost on a window that we can sense is absent in the standard shapes we learn about as children. Simple geometric shapes look different if we change the scale at which we examine them. Consider a circle. As we zoom into its edge, it looks increasingly straight.

Now consider instead a shape formed by iteration after iteration of branching, taking a single line and splitting it into two, then each of these two into two more, and so on.

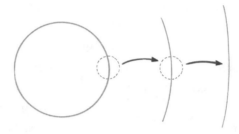

The final shape, after infinitely many levels of branching, looks the same no matter how much we zoom in. It is "self-similar."

Branching is just one of many ways to create self-similarity. A rich area of mathematics called *fractal geometry* examines the properties of self-similar shapes—for example, how they can be generated by processes like diffusion, how they can fill a given space despite a sparsity of substance, and how they behave not as one-, two-, or three-dimensional objects but rather as objects with fractional dimensions (hence the name *fractal*). Many features of the natural world are well described by fractals. Fine twigs growing from a thin branch look roughly like thin branches growing from a thick branch, which look roughly like thick branches growing off the trunk of a tree; an approximate self-similarity connects features at different scales. As mathematician Benoit B. Mandelbrot noted, "Clouds are not spheres, mountains are not cones, coastlines are not circles." In other words, many of the shapes that exemplify nature are fundamentally different from simple, standard geometric forms. Their self-similar character isn't just a minor modification to spherical, conical, or circular shapes, but is their inescapable essence. Mandelbrot pioneered the analysis of such shapes and coined the term *fractal*. (His middle initial, B., the joke goes, stands for "Benoit B. Mandelbrot.") Though self-similarity of real systems doesn't proceed to infinitely fine scales, fractals nonetheless provide a powerful framework for understanding nature. The exact pathways of blood vessels may differ in their details from animal to animal, but all can be thought of abstractly as fractal branching networks.

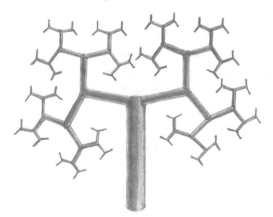

West, Brown, and Enquist recast the mystery of metabolic scaling into a question about fractal networks of pipes. These networks might supply oxygen, blood, nutrients—whatever. Rather than asking what metabolic rate an animal or plant needs, West and colleagues boldly asked what flow its supply network could sustain. This, in turn, is a question of physics and geometry. It's harder to push fluid through a narrow pipe than a wide one. More precisely, a pipe with half the cross-sectional area of a larger pipe needs four times as much pressure to drive fluid at the same flow rate. Branching into ever-finer pipes, as in a network of blood vessels, leads to an ever-greater energetic tax on the organism, the magnitude of which depends on the geometry of the array of tubes. A larger animal needs more levels of branching to connect the cellular scale with the whole-body scale. West, Brown, and Enquist expressed this mathematically and argued that the most efficient self-similar network would scale with body size in such a way that an exponent of ¾ naturally emerges for the scaling of metabolic rate with mass, that is, Kleiber's law.

West, Brown, and Enquist's paper, published in 1997 in the prestigious journal *Science*, made an enormous splash. It has been cited by over 4000 research papers since it was written. Finally, it seemed, we had uncovered the biophysics underlying Kleiber's law.

Sadly, however, the story doesn't end so simply. Determining the energetically optimal network geometry is mathematically challenging, and several researchers found subtle but important flaws in West and colleagues' proof. Just as problematically, the fractal supply model rests on several underlying assumptions—related, for example, to details of the branching architecture—that are not necessarily true. (We don't know enough to know whether they are true or not.) We're still left with a fascinating model, but rather than one that can proclaim Kleiber's law emerges from well-understood basic principles, it's one that tells us Kleiber's law is probably equivalent to a collection of other possible laws of nature, encouraging us to investigate their validity.

Perhaps the greatest contribution of West, Brown, and Enquist's work was that it reinvigorated our thinking about metabolism and spurred many biologists, physicists, and mathematicians to turn their attention to the topic. Some of the thousands of subsequent papers were celebratory. Some were critical. Some applied fractal models to other systems— more on that in a moment. Some pushed the theory in surprising directions. In 1999, for example, Jayanth Banavar (currently a colleague of mine at the University of Oregon), Amos Maritan, and Andrea Rinaldo developed a mathematical model that doesn't require self-similarity at all. They examined the relationship between the rate at which a branched delivery network can deliver material and the volume occupied by the network itself, and argued that the most compact network is the one for which the delivery rate scales as the ¾ power of the animal's total volume (or mass). The idea, then, is that animals evolve to minimize the volume required by their circulatory systems, thereby giving Kleiber's law. Applying this model to animal metabolism again requires assumptions whose validity several researchers have questioned, but the ideas are nonetheless intriguing.

Where do we go from here? How can we understand Kleiber's law and, more broadly, understand metabolism? As mentioned, one possibility is that Kleiber's law isn't really a law and that the ¾ exponent requires wishful thinking to believe in. Another, also mentioned, is that a rough ¾ law is a signature of aspects of shape and form that we still don't understand. Yet another is that Kleiber's law isn't a fundamental aspect of how metabolism must work, but rather emerges as a consequence of the many other biophysical laws that organisms are subject to. For example, imagine that animals were made only of muscle and bone, with a kilogram of muscle and a kilogram of bone, each having its own characteristic and constant fuel requirements, but with the fuel consumption rate being greater for muscle than bone. If the proportion of muscle and bone remains the same regardless of overall size, the animal's overall metabolic rate will simply be proportional to its total mass. However, we saw in chapter 10 that bigger animals have bulkier bones. The overall proportion of bone mass is greater for larger animals to help compensate for the weaker scaling of bone strength compared to weight. The animal's overall metabolic rate would therefore scale more weakly than mass to the first power, giving an exponent below 1.0 for reasons of bone mechanics rather than anything intrinsically related to energy consumption or nutrient use.

Of course, animals are more than just muscle and bone. Every organ and tissue has its own signatures of size and shape, currently understood to varying degrees. Together, these might all intersect to give the metabolic characteristics of organisms, described by simple exponents or perhaps more complex relationships. This perspective has been nicely described by biochemist and author Nick Lane and was precisely formulated by Peter Hochachka's group at the University of British Columbia, Canada. Unfortunately, Hochachka and colleagues' initial model had significant mathematical flaws, and it remains to be seen if this approach can be molded into a rigorous and convincing theory.

WHO CARES ABOUT METABOLIC SCALING?

All we've concluded in this chapter is that we can't conclude anything. You, dear reader, may be wondering why I've dragged you through all this if not to reach a satisfying end. Or you might wonder why *anyone* cares about a puzzle that has so steadfastly resisted solutions. In part, the simplicity of the question—How does metabolic rate depend on mass?—and the hints from data of universal laws are the main drivers of interest. Beyond this, though, the solution matters to more than questions of animal energetics.

How much forest does an elephant need? The answer is set in part by the foliage it must graze on to sustain its metabolism. Shifts in species abundances and changes in available space—driven, for example, by disease, poaching, or deforestation—can ripple through an ecosystem in ways that depend on the metabolic needs of larger or smaller creatures. More grandly, one can ask what the metabolism of the whole ecosystem is and whether there are rules that transcend the details of particular constituent creatures. Brown, West, and others have argued for a "metabolic theory of ecology" along these lines. One might hope that a better understanding of metabolic scaling will lead to better strategies for conservation and land use.

Metabolism influences fundamental features of our existence, such as life span. Our elephant from early in this chapter lives much longer than the mouse. The life expectancy of a typical African elephant is over 60 years. The average mouse, even if it escapes being eaten, lives about 2½ years. The contrast represents a general trend: life span on average increases with body mass throughout the animal kingdom. A plot of life span versus mass for mammals shows a rough power law scaling with an exponent of approximately ¼, again for reasons that remain mysterious. The fact that the Kleiber's law exponent of ¾ is a whole number multiple of ¼ may reflect a shared origin, as the network models imply, or it may be coincidental. The elephant also differs from the mouse in its heart rate; the elephant's heart beats about 30 times per minute, the mouse's about 600. Again, this illustrates a general trend of heart rate slowing with size, and yet again we find a scaling exponent, −¼, that's a multiple of ¼. (The negative exponent means that the relationship slopes downward on logarithmic axes.) The opposite exponents of heart rate and life span together have an intriguing consequence. Multiplying heart rate and life span—beats per minute times the total number of minutes lived—gives the total number of beats the heart performs over the animal's existence. The +¼ and −¼ cancel to give a number of heartbeats that is constant regardless of body mass. (More precisely, $mass^{1/4} \times mass^{-1/4} = mass^0$, a flat line with no slope at all.) The mouse, the elephant, and all mammals great and small each get about a billion heartbeats allotted to them. Humans are an anomaly, by the way. Our life span is about twice what's normal for an animal of our mass—a typical 100 kg mammal lives about 30 years—so we've got about two billion beats at our disposal. These relationships regarding heart rate, life span, metabolism, and more might all be coincidental, or they might point to connections that, if we understood them, could give us insights into some of the fundamental aspects of being alive.

Perhaps most surprisingly, understanding nature's metabolic scaling laws might help us understand unnatural sorts of scaling, such as the activity of cities. Geoffrey West and others, building on the fractal model sketched above, have argued that cities, like living organisms, are governed by the properties of networks. These may be meshes of highways, roads, and lanes; arrays of pipes and wires moving liquid and electrical currents; webs of financial transactions; or social networks connecting people. The data suggest that unexpected scaling laws might govern the dependences of income, road coverage, patent filing, and more on city size. The data are, however, noisy, and their trends are even more contentious than those of animal metabolism. Nonetheless, the possibility of deep principles underlying urban life is tantalizing, especially because the continuing urbanization of the earth's human population calls for better design and management of cities.

As fascinating as the implications of understanding metabolic scaling may be, you might still be frustrated by the present state of confusion. That's fine—I am, too. A fundamental problem is the vast imbalance between abundant theory and sparse data. Papers written about Kleiber's law outnumber the measured data points! Science is much more than theorizing. In fact, what makes science science is that ideas are tested by their ability to make predictions about the real world. There is no shortage of theories that are mathematically sound, elegant, and wrong. I run a research lab, performing experiments rather than working solely with equations and computers, in large part because I want to see, and be surprised by, the way nature actually behaves.

It's hard to experimentally test metabolic scaling ideas. Though we've imagined several times so far animals getting smaller or larger, real elephants and mice don't come with knobs to dial their masses up and down. There are a few clever approaches that get at this obliquely. Frank Jülicher, Jochen Rink, and colleagues in Dresden, Germany, for example, recently examined flatworms that grow or shrink dramatically depending on the availability of food, with a mass range spanning more than a factor of 1000. Strikingly, their measured metabolic rates scaled as body mass to the ¾ power, just as Kleiber's law would predict, and in contradiction to the ⅔ exponent noted earlier that typically characterizes animals of the same species. The researchers were able to attribute this scaling form to mass-dependent changes in how the animals' cells store fats and sugars. Whether this is a general path to Kleiber's law or a peculiarity of flatworms is still to be determined. Our mystery remains mysterious. Its resolution awaits new data, or perhaps new animals.

The scaling relationships we've encountered throughout the past few chapters reflect connections between the cellular machineries encoded in genomes and the large-scale constraints of physical laws. In part III, we see how and why to read and write genomes, potentially redesigning organisms in ways that influence both small- and large-scale function. We might imagine that understanding biophysical scaling relationships enhances our ability to make genetic predictions, or that the outcomes of genetic engineering inform our understanding of scaling laws. Both are likely to be true.

PART III

Organisms by Design

13

How We Read DNA

In the preceding chapters, we've seen that life involves the interplay of physical objects and physical laws—molecules, cells, and organs governed by principles of self-assembly, regulation, randomness, and scaling. The two categories of objects and laws aren't really separate. A protein is a protein, for example, because of the self-assembly of its constituent amino acids, driven by the incessant dance of Brownian motion to explore the possibilities of three-dimensional shapes. An organism's genome ties together tangible stuff and the forces that shape the stuff, encoding the amino acid sequences and regulatory motifs that physical interactions will mold. Changing the genome, therefore, changes the resulting organism.

So far in this book, my aim has been primarily to illuminate the intersection of physical principles and the natural workings of life. Now in part III, we dive into the application of our newfound knowledge to alter how living things function, in ways both subtle and radical. The physical nature of biological matter is crucial not only to the implementation of new technologies but also to their implications. What we infer from differences in DNA sequences, for example, or what outcomes are possible or impossible if we alter such sequences, depends on the processes that orchestrate the readout of genes, the architecture of the proteins formed within our cells, the forces that guide and constrain self-assembly, the randomness inherent in the microscopic environment, and other such biophysical concerns. Keeping in mind the physical context of life will help us form a realistic vision of the impacts of present and future biotechnologies, distinguishing between possibilities that are likely or unlikely, feasible or far-fetched.

In chapter 15, our explorations lead us to methods for rewriting genomes. Before writing, however, we must be able to read, discerning the sequence of As, Cs, Gs, and Ts that defines a given set of DNA. We have achieved this ability, creating dazzling technologies that harness the physical characteristics of this vital molecule and the constructive power of self-assembly and microscopic randomness.

WHY IS DNA HARD TO READ?

Reading DNA in itself provides deep insights into life with thought-provoking, practical consequences. For example, detecting unusual sequences (i.e., mutations) that correspond to increased likelihoods of disorders such as cancer may spur preventive measures. Relatedly, sequencing the cells within cancerous tumors can reveal particular genetic signatures that suggest particular treatments. These applications and many others require maps of a 2-nanometer-wide, 1-meter-long molecule.

We know from chapter 1 that DNA is a chain-like molecule, composed of just four possible units. Why, then, isn't it simple to read its sequence? If I write the word *molecule* on a page, you can immediately see that its sequence is M-O-L-E-C-U-L-E. The problem with DNA is that it's small. Each nucleotide is about one-third of a nanometer long, a nanometer being one-billionth, or 10^{-9}, of a meter. Not only is that small on a human scale, it's too small for any microscope that uses light. Light, like radio waves or X-rays, is an electromagnetic wave, a traveling undulation of electric and magnetic fields. For visible light, the wavelength—the distance between the peaks—is a few hundred nanometers, the exact value depending on the color of the light. The laws of optics dictate that the smallest features one can discern are about the size of the wavelength of light one uses to see them, no matter what lenses, mirrors, or microscopes one builds. Any features finer than this are blurred together. Even if we were to label each A, C, G, and T with a different color or its own distinct flag, our image would blur together each nucleotide's marker with those of a thousand of its neighbors. Trying to simply read out the sequence would be hopeless.

You might guess that the way out of this is to use something other than visible light to form an image, but this approach fails. Smaller waves certainly exist. X-rays, for example, have wavelengths between about 0.01 and 10 nanometers. Electrons behave as waves, and when guided in electron microscopes have wavelengths that are tenths of nanometers or smaller. DNA nucleotides would in principle be resolvable with either of these probes, but multiple problems pop up in practice: X-rays are challenging to focus; the high energies of X-rays and electron beams can be destructive; and the different DNA nucleotides are almost identical from the perspective of X-rays and electrons, so we couldn't determine the sequence even if we were able to form an image. You might find it odd that X-rays aren't helpful, since we learned in chapter 1 that these waves were used to determine the double-helical structure of DNA. This, however, involved shining X-rays through a whole crystal of DNA, a grid of many trillions of identical molecules. The interaction of the waves with all these DNA strands reveals the twisted ladder structure they all share; it doesn't discern the sequence of any individual strand.

PIECING WORDS TOGETHER

Reading DNA, therefore, is difficult. Fifteen years after the 1953 revelation of the double helix structure of DNA, Ray Wu and Dale Kaiser managed to decipher 12 of the nucleotides of a viral genome. Its full genome contains about 48,000 nucleotides. Five years later, Allan Maxam and Walter Gilbert determined the sequence of 24 nucleotides that make up the DNA binding site of the lac repressor (chapter 4). This took two years of demanding work, a rate that would translate to 250 million years for sequencing the full human genome. Clearly, one would hope for better methods.

Better methods emerged, and over the next several pages we look at them in some detail, not only because of their importance to the modern world but also because their very existence is a testament to the power of studying the physicality of DNA and other biological molecules. Properties like size, stiffness, and electrical charge, and themes like self-assembly and randomness, serve as the backdrop of this very practically focused chapter.

Two clever approaches to determining DNA sequences arose around 1977, one by the same Maxam and Gilbert, and the other by Frederick Sanger and colleagues. Sanger's method was the easier of the two and soon came to dominate DNA sequencing, becoming the method of choice for over two decades. I begin our description of how to read DNA with Sanger sequencing.

Imagine that, rather than reading one by one the letters M-O-L-E-C-U-L-E, you instead had many truncated copies of the word in which you could only make out the last letter: "??L," "?O," "?????U," and so on. Noticing that all the three-letter fragments end in L, all four-letter fragments in E, and so on, you deduce that the complete word is MOLECULE. Conceptually, this is the essence of Sanger's method as well as some of the others—reimagining the task of reading as an identification of distinct pieces rather than progression along a strand.

We've already encountered, in chapter 1, several of the steps in the recipe for this approach. Imagine a fragment of DNA—not a whole genome, but something more tractable, perhaps a few hundred nucleotides long. The *polymerase chain reaction* (PCR) generates countless copies of the fragment, and heat separates the two halves of each double helix. Ignore one of these halves for now; imagine millions of identical, single-stranded DNAs. Keep in mind that DNA polymerase generates a perfect complement to any single-stranded DNA by stitching free nucleotides together to synthesize a new partner strand. Now for the new idea: Imagine that the researcher again uses DNA polymerase to replicate DNA as in normal PCR, but spikes the stock of free nucleotides with a small fraction of defective units—slightly different As, Cs, Gs, and Ts that can still be attached to the strand, but to which new nucleotides can't be linked. Each nucleotide addition is a roll of the dice: If DNA polymerase latches onto a normal free nucleotide, the elongation of the strand progresses. If it grabs an altered nucleotide, the elongation terminates with this final unit. Because the addition of terminal nucleotides is random and rare, the researcher ends up with many DNA strands all beginning with the same starting point but extending different lengths.

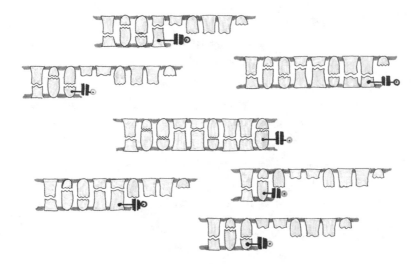

So far, it seems like the consequence of the altered nucleotides is just to botch PCR, but these terminal units are designed not just to thwart DNA elongation but also to emit light—a different color for the not-quite-A, not-quite-C, not-quite-G, and not-quite-T. There are other possible markers than color. Sanger sequencing at first used radioactive labels and didn't use PCR (which hadn't been invented yet), but rather employed bacteria to clone DNA. Our description here is of its later, more efficient variants, but the principles are the same.

Continuing, a final melting step separates each DNA molecule into single strands, with the variable-length fragments labeled at their ends, like our pieces of words with the last letter visible. The researcher still doesn't know how long a given fragment is, however, and the fragments are too small to observe with visible light. As we saw in chapter 1, PCR makes use of one of DNA's important physical properties: its melting, that is, its separation into single strands above some critical temperature. Sanger sequencing builds on this by exploiting another physical property of DNA: its electrical charge.

DNA has a negative electrical charge, as we noted in chapter 3 when discussing its wrapping around histones. As a result, one can move it with electric fields, dragging the molecule toward positive electrodes and away from negative ones. In plain water, pieces of DNA move with similar speeds regardless of their size. Larger fragments have more charge and hence a greater electrical force pushing them along, but also experience more drag from the fluid. The physics governing the scaling of force and drag with fragment size turn out to be complex and subtle, but the ultimate result is a near cancellation of effects that make mobility roughly independent of length in a simple liquid. The situation changes, however, in a slab of gel. In a gel, like the edible gelatin of Jell-O, long, chain-like molecules are pinned to one another, forming a porous, three-dimensional meshwork permeated by water. The single-stranded DNA must snake through the pores to travel through the gel—the technical term is *reptation*—which requires a series of contortions that take considerably more time for longer DNA strands than for shorter ones.

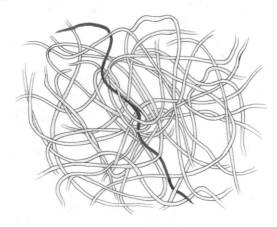

The predictable randomness of Brownian motion is crucial to this transport. Without it, DNA would get stuck in the gel no matter how strong the electric field; if the two ends of a strand fell into different pores, the molecule would be draped across the barrier like a towel on a clothesline, never to be free. Thanks to Brownian motion, however, the DNA is constantly jiggling and reorienting, coming loose from one hole to pass through another. The statistical predictability of microscopic randomness gives a well-defined, and mathematically tractable, speed to the molecule's travel.

The final result of the duplication, end-labeling, electric-field-driving, and dragging-through-gel of the DNA is a straightforward readout of nucleotide sequence. All the fragments of a given length have the same color labeling their end. Strands that are 27 nucleotides long end with C and are red, let's say, with a terminal-modified C. Likewise, 28-nucleotide-long strands ending with T are blue, let's say, with a terminal-modified T. And so on. None of the 27-nucleotide-long strands are blue, since all fragments of that length, being clones of one another, must end in C and all terminating Cs are red. (If you've conscientiously worried for the past few pages about the ignored half of the original double-stranded DNA, which seems like it would contribute another set of molecular fragments, fear not: a judicious choice of primer guides DNA polymerase to work on just one of the single strands in Sanger sequencing, so the other is not replicated at all.)

All the DNA fragments start out moving through a thin tube of gel together, but they separate as they travel at different speeds. Watching a particular point along the tube, the researcher sees a red pulse go by, then a blue one, then another blue one, then a green one, and so on, from which it follows that our original piece of DNA had the sequence . . . CTTA. . . . We have read our DNA.

Enhanced by a proliferation of technological improvements, Sanger sequencing and its variants, often referred to in retrospect as first-generation sequencing methods, were able to read around 1000 nucleotides per day by the mid-1980s. It's more common to see this written as 1000 bases per day and to refer to the length of genomes in terms of the number of bases. For our purposes, the distinction between nucleotide and bases is irrelevant. (If we want to be pedantic, an A, C, G, or T nucleotide consists of an adenine, cytosine, guanine, or thymine base joined to a sugar called *ribose* and a collection of phosphorus and oxygen atoms called a *phosphate group*. *Nucleotide*, in other words, means base + ribose + phosphate, and it's the ribose and phosphate groups that are linked together to make a strand of DNA.)

The sequence of an entire genome is revealed by stitching together all the sequence fragments. In 1982, we had the complete 48,000-base genome of the bacteria-infecting virus that Wu and Gilbert caught a glimpse of in 1968. The full genome of the yeast *S. cerevisiae* (12 million bases) was mapped in 1996 and the nematode worm *C. elegans* (100 million bases) in 1998. The most tantalizing target, of course, was *Homo sapiens*. Though it was clear that Sanger sequencing could in principle tackle the human genome, applying it to a genome billions of bases in size was in practice a massive technological challenge. The task called not only for improvements in biochemistry, related to terminal nucleotides, for example, but for advances in our tools for physical manipulation of DNA—melting, pushing, detecting emitted light, and more, all to be performed robustly and rapidly.

In 1988, the US Congress authorized funds for what became known as the Human Genome Project, set to launch in 1990 with an estimated time to completion of 15 years and an estimated cost of $3 billion. (For scale, total annual federal non-defense-related research spending in the United States in 1990 was about $23 billion.) Much like the Apollo moon program of the 1960s, the Human Genome Project evoked notions of exploring new frontiers, this time of the inner world within cells rather than outer space. The Human Genome Project was publicly funded, organized mostly through the US National Institutes of Health and the Department of Energy, though with considerable international collaboration. In 1998, however, a privately funded group led by biotechnologist Craig Venter announced a plan to independently sequence the human genome, and moreover to do so faster and at a lower cost than the Human Genome Project. A race was on! Both groups succeeded and in 2001 jointly announced a 90% complete version of the human genetic code. In 2003, they reached 99% coverage, declaring the achievement of an essentially complete genome two years ahead of the original schedule. Work continued to fill in the few remaining segments that defied initial efforts due to difficulties such as repetitive sequences, and by 2004 they had a 99.7% complete genome.

You might wonder, *Whose* genetic code was sequenced? Both projects used composites: DNA from several individuals so that some fragments came from one person, some from another, in total giving a generic picture. In practice, however, the majority of the genetic material in each project came from a single individual. For the Human Genome Project, this was an anonymous man probably from Buffalo, New York, and for Venter's effort, this was an anonymous donor later revealed to be . . . Craig Venter himself! These people are certainly not representative of the whole of humanity; to understand our species, we'd like a statistical portrait, which would require sequencing many more humans' genomes. Similarly, if I were to develop cancer, my doctor would like the genome of *my* malignant cells, not a generic average. Addressing these limitations calls for much faster and cheaper technologies. Luckily, these were just around the corner.

READING MANY WORDS AT ONCE

The $3 billion price tag of the Human Genome Project works out to about $1 per base pair. This is a stunning achievement given that less than a human lifetime earlier we didn't even know the structure of DNA; nonetheless, it's far too expensive for routine applications. In the early twenty-first century, several clever new techniques emerged, spurred in part by public funding for sequencing innovations. These approaches are known collectively as next-generation, second-generation, or high-throughput methods. In Sanger (i.e., first-generation) sequencing, one replicated fragment is sequenced at a time. Mixing fragments together would be disastrous, as we'd lose the unique correspondence between the length of a truncated subfragment and its terminal nucleotide. Second-generation methods are inherently parallel, capable of assessing many fragments at once and in many cases reading DNA strands as they grow. We look at a few of these second-generation sequencing approaches now. Though diverse in their details, all these methods share a theme: leveraging physical attributes of DNA, DNA-associated materials, or both.

Pyrosequencing is a next-generation method made possible in part by the awesome abilities of fireflies. As we've noted, DNA polymerase stitches new nucleotides onto DNA strands. A careful accounting of the atoms that make up the nucleotide and the atoms that make up the strand shows that they don't quite match up. The reaction that links a nucleotide to a DNA chain liberates a tiny molecule called *pyrophosphate*, made of two phosphorus atoms and seven oxygen atoms. A protein included in the pyrosequencing soup transforms pyrophosphate into ATP, a molecule used by cells for many different activities. One such activity is a light-emitting chemical reaction performed by proteins called *luciferases* that use ATP as their fuel. (The Latin *lucifer* means "bringer of light.") Creatures such as fireflies, click beetles, and luminous fungi naturally make luciferases. As we saw with jellyfish-crafted green fluorescent protein in chapter 2, the variety of life provides a variety of tools that we can repurpose in creative ways.

Pyrosequencing therefore works as follows: As with Sanger sequencing, one starts with DNA fragments, replicated to give many copies and melted into single strands. Again, DNA polymerase will grow the second, complementary strand from a given single strand. Imagine for the moment that there is only one molecule of single-stranded DNA, anchored to a beaker. The researcher pours in a soup that contains luciferase and the other ingredients but that includes only one type of nucleotide—A, let's say. Detecting a pulse of light implies that A must have been added onto the strand by DNA polymerase, and so A must be the next nucleotide in the DNA sequence. Darkness, in contrast, implies that A wasn't added and isn't the next nucleotide. The researcher removes the A-containing soup and repeats the process three more times, with C, G, and T. One and only one of the four will yield a burst of light. The letter is now known. Repeating the process gives the next letter, the next, and so on. The DNA is read as it's being formed.

I haven't explained how this procedure could be parallelized. Moreover, it seems to require perfect sensitivity—production of a single pyrophosphate must lead irrevocably to a single luciferase emitting a single, exceptionally faint pulse of light that we must detect perfectly. If any of these steps fail, we miss a letter of our genetic code. These two issues, parallelism and robustness, are addressed by the same physical tactic: bunching many identical DNA fragments together.

As in Sanger sequencing, one first fragments DNA into small pieces, less than 1000 bases long. The DNA is melted to separate the two strands of each double-stranded fragment (chapter 1). A short sequence of chemically modified bases links each single-stranded DNA to the surface of a microscopic bead, with DNA and beads mixed together with such an excess of beads that it is very unlikely that any bead will have more than one DNA molecule bound to its surface.

The beads and DNA are in a watery solution. Mixing with oil under agitation or flow gives water droplets not much larger than the beads, encapsulating no more than one bead within.

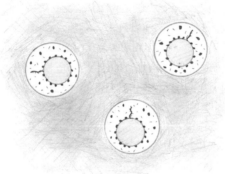

The surface of the beads is decorated with primers to initiate DNA polymerase's DNA synthesis. The watery solution contains DNA polymerase and nucleotides, as well as primers that enable replication of the single strand's complement. Each droplet therefore holds all the ingredients to generate vast numbers of copies, around a million typically, of the starting fragment. After the replication is done, one collects the droplets and adds soap or alcohol to lower the surface tension that keeps each drop isolated in oil (chapter 11). The water coalesces and flows over a plate dotted with tiny wells just larger than a single bead, so that each well is home to just one DNA-coated sphere. At this point the

DNA is double-stranded, with one of the strands held by the surface-attached primers. Melting the double-stranded DNA and washing away the unbound strands leaves an array of tiny globes, each covered with its own forest of identical single-stranded DNA molecules.

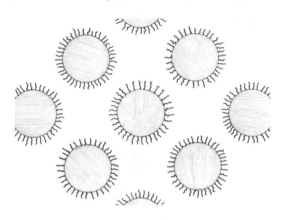

Now pyrosequencing can begin, watching for flashes of light in each well as the million-fold-replicated DNA grows, a million times brighter than would be the case for a single DNA molecule. If there are repeating letters in the DNA sequence, the pulses will be brighter, twice as bright for two As in a row, for example, as for one. The researchers, or more accurately their machines, tabulate the sequence of colors and intensities, and thereby read the DNA.

This elaborate scheme actually works, and is robust enough to have been the first next-generation sequencing method to be commercialized, by a company called 454 Life Sciences in 2005. For about $500,000, you could buy a machine to perform pyrosequencing and output the nucleotide letters making up the DNA input. At half a million dollars, you wouldn't buy one for your living room, but it was well within the grasp of moderately sized research institutes. (For scale, the median house price in the United States in 2005 was about $240,000.) Pyrosequencing machines are no longer sold, however, superseded by other sequencing technologies that are no less amazing.

A similar method to pyrosequencing makes use of another piece that's left over from the stitching of a nucleotide onto a DNA strand. This piece is so small and seemingly insignificant that its detection is especially astounding. It's a single proton. All the everyday matter surrounding us is made up of protons, neutrons, and electrons. The lightest element, hydrogen, is simply one proton, positively charged, and one electron, negatively charged, bound together. A lone proton isn't even that. Detecting it quickly and robustly is made possible by a very nonbiological technology: the transistor.

Transistors make up the circuitry of cell phones, computers, and countless other electronic devices. In every transistor, the flow of electrical current from one point to another is governed by what happens at a third point, a bit like boat traffic along a river being governed by a drawbridge operator. In so-called field effect transistors, the controlling factor is an electric field, for example, from a proton in proximity to the transistor surface.

As in pyrosequencing, beads coated with cloned DNA fragments settle into individual wells. The array of wells sits atop a semiconductor chip, constructed so that each well has a field effect transistor at its base. Rather than pulses of light, one detects pulses of electrical current. This technology was commercialized by Ion Torrent, a company founded by prolific biotechnologist Jonathan Rothberg, who previously headed 454 Life Sciences. Ion Torrent introduced its Personal Genome Machine in 2010. It fit on a tabletop and cost just $50,000, less than twice the average price of a new car sold that year.

The dominant next-generation sequencing technology, however, made use of clever chemical modification of DNA rather than sophisticated detection of nucleotide addition. Recall that in Sanger sequencing the growth of a DNA strand is terminated with an unnatural nucleotide. By the late 1990s, more flexible methods were in hand: reversibly terminating, reversibly fluorescent nucleotides. One at a time, the researcher adds to replicated DNA fragments a modified A, C, G, or T, each with a different color. Washing away the free nucleotides reveals the color corresponding to the one that was glued by DNA polymerase onto the next position of the growing DNA. The fluorescent part of the molecule and the part that blocks DNA polymerase activity are both linked to the normal nucleotide part by a strand of atoms that can be chemically cleaved. Flowing in the cleaving agents leaves regular, unadorned DNA, ready for the next nucleotide addition; the process repeats. Typically, the process is performed using glass slides dotted with patches of replicated DNA fragments, with cameras capturing the flashes of color appearing and disappearing at each spot on the slide.

This technology is commonly known as *Illumina sequencing*, after the company that acquired its original developer. From them, one can buy instruments at prices from about $100,000 to millions of dollars, depending on parameters like how many nucleotide bases are read per run. As with all sequencing technologies, it takes time and money to prepare the DNA sample and assemble the chip with wells, the slide with patches of DNA clones, or some other platform. Apart from purchasing the sequencing machine, the cost to sequence a billion DNA bases ranges from about $5 to $150 for Illumina sequencing, in contrast to about $10,000 for semiconductor sequencing, contributing in large part to Illumina's appeal. I comment more on price shortly, but it's worth noting that all these numbers are far smaller than the $3 billion it cost to read three billion bases in the initial Human Genome Project!

SINGLE WORDS FROM SINGLE MOLECULES

It's harder to identify the boundary between second- and third-generation sequencing methods than that between first- and second-. The invention of new techniques doesn't come in bursts separated by periods of rest, but rather by continuous activity along multiple overlapping fronts. Nonetheless, a convenient and roughly chronologically correct distinction is that third-generation methods involve the sequencing of *single* DNA molecules, without the need for replicas. In an approach commercialized by a company called Pacific Biosciences, for example, individual DNA polymerase molecules are anchored in tiny pits carved into a metal film, the optical properties of which dictate that fluorescent nucleotides are visible only while being stitched onto the growing DNA molecule.

The most remarkable new DNA sequencing technology doesn't involve growing DNA, colorful nucleotides, or even DNA polymerase. We could ignore everything we've learned over the past few pages about methods in which sequencing is linked to synthesis, and return to the naive picture of simply proceeding base by base along an already formed DNA strand and recording what's there. As noted, one can't use light to distinguish bases. One can, however, use electricity.

Imagine a pore in a membrane that separates two liquid reservoirs. The watery solution on each side contains ions, which are electrically charged atoms or molecules. A voltage applied across the membrane causes ions to flow through the pore, and one can measure this electrical current. If the pore is partially blocked, the current will be smaller; the magnitude of the current reflects the accessibility of the pore to ions. Now imagine a single strand of DNA threading the pore. Each base, A, C, G, or T, has a different number of atoms and a different physical size, and therefore allows a different current flow if it is situated in the pore. To turn this into a sequencing method, one needs to drag the DNA through the hole. In early setups, the voltage difference was also the driver of DNA motion; as we've seen, DNA itself is electrically charged. This works poorly, however, mainly because the transport is too fast to allow precise current measurements for each base passing by. The solution to this problem? Proteins. There are many different proteins that ratchet themselves along a DNA strand as part of their normal function—DNA polymerase, for example. Fixing one of these proteins to the pore (green in the illustration below), the now stationary protein's ratcheting equipment feeds DNA through the

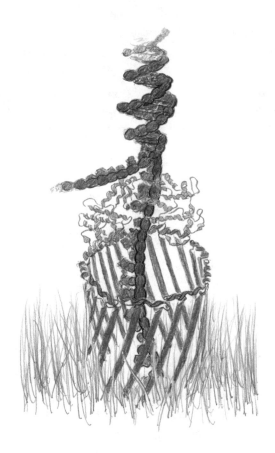

opening. We saw in chapter 2 that many proteins are natural nanometer-scale machines. Here that machinery is harnessed to perform work in an otherwise inaccessible space. Proteins can also form the pores themselves, employing molecules like those we saw in chapter 2 that fold into shapes with channels running through them. Once again, we see the manifestations of self-assembly.

This *nanopore sequencing* scheme is conceptually simple and has been hypothesized since the 1980s as potentially viable. It took a lot of work to actually realize it, however. Researchers first demonstrated the reliability of pore insertion into artificial barriers, not with the aim of DNA sequencing but simply to understand pore structure and develop platforms for working with membranes and proteins together. Early studies of electric-field-driven DNA motion through pores helped researchers understand the physics underlying transport in constrained spaces. Even with the principles in place by the early 2000s, many casual observers, myself included, found it hard to imagine that nanopore sequencing would become a robust, reliable technology. One must put together pores, DNA-ratcheting proteins, few-nanometer-thick membranes, and electrodes. Then one must perform exquisitely sensitive current measurements while ensuring that the pores don't clog due to contaminants or damaged proteins. This daunting engineering challenge was taken up by Oxford Nanopore, a company focused, as you'd guess from the name, on nanopore-based detection of molecules. Its first instrument was released to researchers in 2014 as a pilot program to evaluate its efficacy, and a few different machines are now available commercially. The smallest is thumb-sized, like a USB memory stick, and in fact plugs into a computer's USB port. A major part of its appeal is portability together with low cost. In 2014, for example, scientists sequenced Ebola virus DNA from 14 human patients in Guinea. They mapped the viral genome within two days after collecting samples, highlighting the potential for rapid sequencing in the field that may enable fast identification of the origin of outbreaks or the discovery of potentially harmful variants as they arise.

The drawback of nanopore sequencing is relatively low accuracy; a few percent of bases will be read incorrectly. Illumina sequencing, in contrast, misidentifies around 0.1%. Whether this is a concern or not depends on the application. For a detailed map of a large genome or prenatal tests for small but significant mutations, minimizing errors is crucial. For surveillance of potentially harmful microbes, or surveys of the membership of bacterial communities, lower accuracy may be a fine trade-off to gain speed and convenience.

THE EVER-CHEAPER GENOME

Advances in DNA sequencing have been stunning, and the future will likely bring even more. I've briefly mentioned the price of these technologies, but now let's look more closely at the cost of sequencing—how much a researcher would pay to determine the specific series of nucleotides making up a DNA sample. A good measure is the cost of sequencing per human genome, in other words, the cost of sequencing three billion bases. Recall the $3 billion cost of the first human genome, corresponding to $1 per base. Even by 2006 this had fallen over 99% to about $13 million. The graph of cost over time, reproduced here, is amazing.

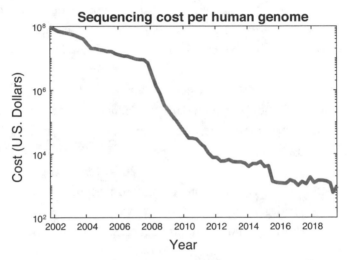

The vertical axis is logarithmic, with divisions being powers of 10; the horizontal axis is not. The decrease in sequencing costs has been strong and sustained. The plunge around 2007, a thousandfold drop in half a decade, marks the transition to next-generation sequencing methods. The time required for sequencing has plummeted along with the cost, from several years for the Human Genome Project to a few days for contemporary instruments.

Perhaps the closest comparably rapid technological advance has been our ability to make transistors. The first transistors were bulky, finicky, and millimeters in size. Now we routinely place billions of transistors on centimeter-square chips, sold for as little as a few dollars apiece. In 1965, Gordon Moore, later CEO of Intel, noted that the number of transistors on a chip doubled every year (or every two years in a revised estimate), an observation that came to be known as Moore's law. Moore's law served as both a description and a manifesto of electronic technology. DNA sequencing, at least as measured by the cost per nucleotide, has advanced at an even more rapid rate. Fittingly, the first human genome to be sequenced using Ion Torrent's Personal Genome Machine was that of Gordon Moore. The drivers of the sequencing revolution are numerous: human ingenuity, the economic incentives of commercialization, public funding for basic research as well as specific programs targeting sequencing innovations, and of course the accumulated understanding of biology, chemistry, and physics that we've managed to amass.

WHEN IS A DNA SEQUENCER
NOT A DNA SEQUENCER?

There's a lot of information we can glean from DNA sequences: differences in the genetic code between species, between individuals, and between cancerous and normal cells, for example. Nonetheless, it may turn out that the biggest impact of DNA sequencing technologies lies not in sequencing genes or genomes at all, but rather in using these tools to reveal the inner lives of cells.

As an example, let's consider RNA. As noted in part I, genes are read by the machinery of RNA polymerase, which transcribes a DNA sequence into an RNA sequence that is then translated into a particular amino acid chain. Ignoring egg and sperm cells and certain immune cells, all the cells in your body have the same genome; sequencing every cell's DNA would be redundant. Knowing what RNA molecules are present in each cell, however, would be interesting—we'd see what genes were on or off as cells developed into blood cells or neurons, or as cells responded to changes in diet or stress. For this we would need to sequence RNA molecules—that is, read their sequence of nucleotides—keeping track of which RNA came from each cell of interest. By now you might guess at aspects of the methodology: harnessing physical forces and properties, with some assistance from biological tools that nature has already developed.

Rather than jellyfish or fireflies, this time our natural aides are viruses. The "central dogma" of biology is that DNA encodes RNA, which encodes proteins. It was a shock, therefore, when the labs of David

Baltimore and Howard Temin in 1970 each independently discovered that certain viruses are capable of the opposite work flow, transcribing their RNA genome into DNA that is inserted into the genomes of their hosts. The protein machine the viruses use is called, appropriately, *reverse transcriptase*, and like DNA polymerase and other such tools, we can co-opt it.

Extracting RNA from a cell, one can add reverse transcriptase and free nucleotides to generate DNA that is complementary to the RNA strands. If the RNA reads "CAGUUGGA," for example, the complement in DNA is "GTCAACCT"—recall that U in RNA is the analog of T in DNA (chapter 3). Subsequently sequencing the DNA reveals its code, and therefore the RNA's code as well. To resolve the RNA library within individual cells, researchers employ a suite of methods similar to those we've already seen, for example, encapsulating single cells along with beads and the requisite biochemical ingredients in vinaigrettes of water and oil. Each RNA molecule is transcribed into DNA, the DNA is sequenced, and we're presented with a readout of what genes were "on" in each cell.

Though single-cell RNA sequencing is a destructive measurement—cells don't survive being burst open—one can apply it to similar cells collected at different times during some process, or with and without some treatment, to reconstruct cellular trajectories. For example, researchers have studied immune cells from organisms exposed, or not, to some pathogenic stimulus to see how gene expression changes in response to provocation. As a second example, RNA from embryos of zebrafish or mice captured at different stages postfertilization reveals the flow of gene expression patterns that guide cells into their roles during animal development.

RNA sequencing is one of many technologies enabled by DNA sequencing. Researchers can now also assess which segments of DNA are wrapped around histones, which have methyl group tags attached, which have transcription factors bound to them, and more. We asked at the top of this section, "When is a DNA sequencer not a DNA sequencer?" The answer? When it's an RNA sequencer, or a DNA packaging mapper, or a guide to gene regulation, or any other such machine.

Living things routinely handle the information in DNA, copying it when cells reproduce and transcribing and translating its sequences into RNA and proteins. The outputs of these processes depend on the sequences of As, Cs, Gs, and Ts, so the processes themselves are, in a sense, reading DNA molecules. For about four billion years, these methods for reading DNA were all that existed. Now we've invented radically new tools— fast, cheap, and almost magical in their efficacy—that place in our sights the information encoded inside every organism. This amazing technological transformation occurred because we took seriously the tangible, physical characteristics of biological molecules, building interfaces between them and other manifestations of our technologies. We turn next to what we can infer from the information encoded in DNA.

Genetic Combinations

The information encoded in all sorts of organisms, including ourselves, now lies before us thanks to our marvelous ability to read DNA. What can it tell us? We asked this question in part I and focused on the nature of genes and gene regulation. It is often tempting to think it straightforward to connect a creature's genetic code with its traits: we merely assess what variations of a gene correspond to variations in the characteristic of interest. Even from our discussions in part I, we know that reality isn't so simple; it isn't just the genes in a genome that dictate biological activity, but also the regulatory circuitry encoded in DNA that switches the transcription of genes on and off. Now we'll see that nature's complexity is even greater than we may have thought; many of the traits and diseases we care about are influenced by the DNA at thousands of sites in the genome, forming a web of connections that is daunting to make sense of.

The theme of randomness introduced earlier comes to our rescue, giving us conceptual and practical tools for handling genetic information. These tools are so effective that we can often get by with maps of the genome that are much sparser, and cheaper to acquire, than the whole genome sequences introduced in the previous chapter. Understanding the meaning of randomness and predictability is crucial for making sense of technologies that already have a large impact on our world, intersecting with topics in health, industry, and ethics in ways that we will touch upon.

WHERE'S THE TALLNESS GENE?

There are many characteristics that fall under the umbrella "genetic," meaning that they're at least influenced, if not wholly determined, by one's sequence of As, Cs, Gs, and Ts—the code we get from our parents. In some cases, including several involving debilitating disorders, there's a simple mapping between what the body does and which span of DNA is ultimately responsible, with just a single gene involved. Cystic fibrosis provides a good example.

We all have mucus lining our lungs, making up the liquid coating we encountered in chapter 11. Cells secrete mucus and drive it along the airways, toward the mouth, sweeping away fluid as well as dirt, pollen, bacteria, and other particles we may have inhaled. The mucus in people suffering from cystic fibrosis is exceptionally viscous, stagnating in the lungs rather than flowing, which leaves the lungs susceptible to bacterial infection. All this is the fault of a single gene that encodes a single protein, the cystic fibrosis transmembrane conductance regulator, which sits in cellular membranes and facilitates the flow of ions into or out of the cells. In cystic fibrosis patients, a mutation in the cystic fibrosis transmembrane conductance regulator gene alters the structure of the protein. As a result, the concentrations of ions on each side of the membrane aren't what they should be, causing water to be sucked out of the mucus, the viscosity of the mucus to increase, and the patient to suffer.

At the other extreme from cystic fibrosis are traits like height. There are nongenetic influences on height—nutrition, for example, plays a major role—but the genetic code you carry from the moment of your conception strongly influences how tall you'll be when you're fully grown. There is not, however, a height gene. Rather, there are tens of thousands of sites in the human genome at which the identity of that site—A, C, G, or T—somehow influences height. These sites aren't all located at genes; many are parts of sequences that regulate gene expression or DNA packing (chapter 3), pulling the strings that guide the genes themselves.

The case of height is much more typical than that of cystic fibrosis, at least for the complex traits and diseases that occupy much of our attention these days. There's no single gene that determines your risk of colon cancer or even your hair color, but rather a polyphony of pieces of the genome. This shouldn't surprise us. After all, with only 20,000 protein-coding genes in the human genome and a complexity far greater than could be described by 20,000 protein-based instructions, it would be astonishingly unlikely that one-to-one mappings between gene and trait would be the norm. The protein encoded by a gene can play roles in many different traits; conversely, a trait can be controlled by the co-ordinated activity of many different genes. Most important of all, as we saw in chapters 3 and 4, the 99% of the genome that doesn't con-sist of genes has a major influence on how genes work, regulating their activation and deactivation.

Let's look more closely at height, both because it's familiar and because it's an excellent example of what genetics can and can't tell us. Height is influenced both by an individual's environment and by genes, and these two driving forces—nurture and nature—are not mu-tually exclusive.

On average, people were shorter in the past. The average Frenchman born in 1800 was 5 feet, 5 inches tall (164 cm); if born in 1980, he was almost 5 feet, 10 inches (176.5 cm) tall. A typical Japanese woman born in 1900 was 4 feet 8 inches tall (143 cm); her descendant born in 1980 was 6 inches (15 cm) taller. The same patterns have been repeated worldwide, especially as countries develop modern economies. These century-or-two spans haven't seen mysterious plagues that wiped out short people, nor striking mutations of the human genome. Rather, the increases in height are driven primarily by nutrition. Our modern Frenchman has about twice the calories per day at his disposal as his turn-of-the-nineteenth-century counterpart. Calories aren't everything, but the abundance of dietary energy correlates with an abundance of nutrients that together allow the human body to develop to its full po-tential. Height is influenced by other nongenetic factors as well, such as childhood diseases and environmental pollutants, exposure to which has decreased markedly for vast swathes of humanity over the past few hundred years.

Now, however, let's consider the population *within* a modern, indus-trialized country. The heights of its adults, even separated by sex, aren't all the same. What's more, we all know that taller parents tend to have taller children. Children tend to grow to heights more similar to their biological parents than to random adults or to adoptive parents. Genetics matters. How much? Which regions of the genome are responsible?

We saw in the last chapter exquisite tools for reading DNA. For rare traits, subtle variations, or comprehensive characterizations, it's useful to sequence entire genomes. Height and many other characteristics, however, turn out to be robust enough that we can use simpler methods. Your genome and mine are over 99% identical, and so we can focus our attention on the regions where we differ. Consider a rare site of disagreement, a location in the genome where most people have an A nucleotide, for example, but a sizable fraction of people have a C nucleotide instead. Such sites of relatively common differences are called *single nucleotide polymorphisms*, abbreviated SNPs and evocatively pronounced "snips." There are a few million common SNPs in the human genome, where "common" means that at least 1% of the population has the rarer nucleotide. A few million is a large number, but it's much smaller than the three billion nucleotides of the full human genome, allowing us to find these SNPs fairly easily. For example, one can coat microscopic beads with short, single-stranded DNA whose sequence includes the complement of the dominant form of an SNP, and then assess whether a test subject's DNA, fragmented and replicated, binds to the beads. Binding indicates that the test sequence matches the SNP; failure to bind indicates a mismatch. I'm glossing over the details and there are several different technologies available, but they're all echoes of the elegant approaches we examined in the last chapter that harness fluorescent nucleotides, DNA polymerase, mass-produced glass slides dotted with millions of beads each with millions of DNA clones, and so on. The end result is that for well under $100 per test, less than many people pay for a pair of shoes, one can determine the ensemble of SNPs that characterize a genome and that therefore characterize much of the genetic variation represented in that individual.

It would be reasonable to guess that knowing one's SNPs would provide little insight, because these sites are only a small fraction of the genome and genomes are complicated. Initially, this seemed to be the case. The first studies mapping SNPs to height, reported in 2008, uncovered about 40 genetic variants that together did a detectable but feeble job of accounting for how tall the people studied were. Even in 2008, it was clear that examining more people would be crucial. The reason is not biology per se, but rather the relationship between randomness and predictability.

As in chapter 6, let's think about flipping coins. Imagine 10 tosses of a fair coin. On average, we'd expect the coin to land heads five times and tails five times, but it wouldn't shock us if we found six heads and four tails. In fact, the probability of this outcome is just 83% that of getting five heads. If you toss the coin 1000 times, 500 heads and 500 tails would be the most probable outcome, and the greater number of flips smooths over the variation so that finding 600 heads and 400 tails becomes far less likely, in fact about *one billion* times less probable than 500 heads. Suppose you suspect that your coin is counterfeit and unbalanced, with a greater than 50% chance of landing heads. If you flip the coin 10 times, you wouldn't be very sensitive to this imbalance—if you found six heads, you wouldn't necessarily conclude that the coin has a 60% chance of landing heads. If 1000 flips gives you 600 heads, however, you'd have a strong indication that your coin is unfair. More precisely, our sensitivity to biased coins scales as the square root of the number of coin flips. This square root may remind you of the statistical properties of our random walker in chapter 6. The connection is more than superficial; the dependencies arise for very similar mathematical reasons.

Returning to the genome, our SNPs are like the coins, and the challenge is to figure out how "fair" or "unfair" each one is—that is, how much each SNP contributes to a trait being different than its average, expected value. An SNP at which the rare variant is equally likely to be found in a tall individual as a short one would be analogous to our fair coin, equally likely to contribute heads or tails. An "unfair" SNP, in contrast, gives a nudge toward being taller or shorter than the average person, like an unfair coin gives a nudge toward the total number of heads being greater or smaller than 50%. The nudges may be very small. The analog to tossing lots of coins is examining lots of people's genomes. To assess the bias of any particular SNP, we can look at the correlation between a person's DNA at that SNP position and the person's height, assessed for as many people as we can. The more genomes we examine, the more sensitive we are to identifying height-related SNPs.

We're now firmly entrenched in the era of large-scale genome studies. A group led by Michigan State University's Stephen Hsu, a physicist by training, examined data from nearly half a million people collected as part of the United Kingdom's Biobank project, detecting from their statistical signatures any SNPs that correlated with the subjects' heights. The researchers found far more than the 40 SNPs of the 2008 study— in fact, about 20,000. Such analyses are challenging and it's easy to be fooled by false patterns. There are mathematical tests to evaluate the reliability of one's assessments, but the best test is to see whether the SNPs one has identified as being correlated with height in the genomes under study work as predictors of height in a different set of people. In other words, one examines most, but not all, of the Biobank dataset, discovering, for example, that SNP number 312 corresponds, on average, to a 0.05 cm increase in height relative to the average; SNP number 3092 corresponds to −0.02 cm; SNP number 4512 corresponds to +0.08 cm; and so on. Next, in the part of the dataset so far left unexamined, one looks at each person's SNPs and adds up the inferred effects, predicting what their height should be. One then assesses how well the prediction matches their actual height. Hsu and colleagues did this in their 2018 paper, finding that most of the measured heights were within an inch (3 cm) of the SNP-based predictions. To get a better sense of what this sort of accuracy looks like, let's make some graphs.

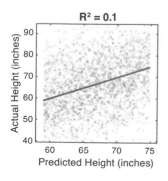
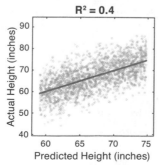
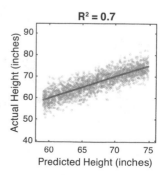

Each graph shows a cloud of hypothetical data points, the predicted heights measured on the horizontal axis and the actual heights on the vertical, with each dot corresponding to one person. The actual and predicted heights are correlated in all three plots. Moreover, the line that best fits the data points is nearly identical in each graph. How well that "best-fit" line actually describes the data, however, is markedly different in the three cases. On the left, the cloud is very diffuse; in the middle, the data points are better localized near the line; on the right, the measured values quite tightly cluster around the prediction. This variation can be quantified by a statistical property called the *coefficient of variation*, often denoted by the symbol R^2. For an intuitive description of R^2, first imagine measuring how scattered the points would be around a flat, horizontal line going through the middle of the graph. (If you know some statistics, imagine the variance of the measured data points.) Next suppose we measure how scattered the points are around our best-fit line. This amount of variation is smaller—it's what's left over after accounting for the relationship revealed by the line. The ratio of this second variance to the first is a number between zero and one, which is smaller the more tightly the points cling to the best-fit line. One minus this ratio is the variance "explained" by the linear relationship; that's R^2. In the leftmost graph, the diffuse cloud, R^2 is 0.1, meaning that the best-fit relationship between prediction and measurement accounts for 10% of the scatter of the data points. In the rightmost graph, R^2 is 0.7; 70% of the scatter is accounted for.

In the SNP-based height analysis of Hsu and colleagues, R^2 is about 0.42, similar to the middle graph—not a perfectly tight spread but not a formless cloud either, and good enough to correspond to the precision of an inch mentioned earlier. An inch may not seem so impressive, but it's more accurate, it turns out, than the prediction you'd make about someone's height by knowing how tall their parents were. And, of course, the SNP-based assessment doesn't require any information about identity or ancestry—just DNA and a cheap test. As Hsu points out, the debris from a crime scene now suffices to tell you how tall a completely unknown person is, and it will reveal a host of other physical characteristics as well.

How good could the R^2 of height get? From studying family members of different levels of relatedness, including identical twins (whose genomes are exactly the same), geneticists have long known that the heritability of height is about 80%. In other words, heredity explains about 80% of the variation between individuals. Is the gap between 0.4 and 0.8 due to the DNA that's inaccessible to SNP studies, or does it arise from more mysterious biological mechanisms? In 2019, Australian geneticist Peter Visscher and colleagues examined whole genome sequences from over 20,000 people and found that the information encoded by DNA does, in fact, explain 80% of the variation in human height. At least for modern Europeans, the vagaries of diet, disease, and exercise give rise to 20% of the spread in the heights we see.

BUILD A BETTER CHICKEN . . .

Of course, all this applies to much more than humans. Rather than the heights of our neighbors, we could just as well ask about the genetic contributions to the spots on a leopard, the petals of a rose, or the mass of an amoeba. Controlling variation in the traits of organisms is central to agriculture. The human population of our planet doubled from two to four billion between 1930 and 1970, and has doubled again since then. That this stunning increase hasn't been met by widespread starvation is due to a remarkable range of innovations. A central feature of the Green Revolution of the 1950s and 1960s, for example, was the selective breeding of new strains of wheat and rice. American agronomist Norman Borlaug, working in Mexico in the mid-twentieth century, bred wheat varieties with large, seed-bearing heads. These plants, however, had a tendency to fall over—as we saw in chapter 10, it's hard to be big. Crossing the plants with dwarf strains—mutants from Japan—gave robust, high-yield wheat. With these and similar advances, Borlaug is considered to have saved a *billion* human lives.

We want shorter wheat but larger chickens. Contemporary chickens raised for consumption in North America weigh four times as much as their ancestors of the 1950s, even if raised on the same feed. (To get a sense of how dramatic this increase has been, picture a world in which the average person weighs 700 pounds (320 kg).) There is variety in chicken body types, some amount of which is due to genetics, and the hulking modern fowl are the result of consistently selecting the larger chickens as those to breed. In humans, by the way, the human genome study of Visscher and colleagues noted above assigns to DNA sequence about 40% of the variation in body mass index, a measure of relative body size and mass.

Nowadays, rather than relying solely on easily identifiable traits, one can use SNPs to help choose which animals or plants to mate. Selecting a large hen and a large rooster, for example, may give large chicks, but what one would really like is for the size-enhancing variants in the mother's genome to be different from the father's, so that the children are more likely to possess two distinct sets of genetic nudges toward being large—two different unfair coins in the chick's pocket. Collecting SNP data, or "SNP genotyping," is therefore increasingly common. In 2019, the US dairy database, for example, contained genotypes from three million cows, up from two million just two years before. Dozens of crops from wheat to tomatoes to sunflowers are covered by SNP tools and databases.

. . . AND A BETTER WATERMELON

As we saw for human height, and as holds for a host of disorders that we'll come to shortly, SNPs provide predictive power despite their sparsity across the genome. This is surprising, and the minimal commentary I gave earlier that perhaps the single nucleotide variation lies at a gene or a regulatory region is perhaps unsatisfying. There's another reason SNPs convey so much meaning that recalls a key point from the first chapter: DNA is a physical object, a long, chain-like molecule. The DNA we received from each of our own two parents is combined in the cells that seed our offspring. In the cell that creates a sperm or egg cell, a strand from one parent and a strand from the other are close together. Protein machines can exchange segments of the strands, cutting and pasting to mix the two genomes, resulting in final strands that are each mixtures of genetic information from the two parents. The randomness of where and whether exchanges happen ensures that each sperm or egg cell is, for all practical purposes, unique, one of a vast number of possible splices. The exchanged DNA segments may include genes, regulatory regions, and SNPs. The closer together any regions are on the parental genome, the more likely they'll travel together in these DNA exchanges and, when sperm and egg meet, be passed on to the resulting child. An SNP, therefore, isn't just an SNP; it can be a marker of a larger expanse of DNA surrounding it that more directly shapes the organism. Sequencing these single nucleotide variants gives us a proxy for more detailed genetic information.

Watermelons illustrate the consequences of physical proximity between genes. Ten thousand years ago, there were no watermelons. The ancestor of the modern plant was a desert vine whose small fruits held pale yellow, bitter flesh. They readily retained water, however—a valuable trait. The ancient Egyptians farmed and domesticated the gourd, selecting generation after generation the seeds that came from the least bitter fruits, generating the sweetness that we continue to love. One of the genes most responsible for taste, whose variants in DNA sequence result in varying degrees of bitterness or sweetness in the fruit, sits close to a gene that encodes a protein that influences color. Selecting for the sweeter variants, therefore, brought redder flesh along for the ride. Continued selective breeding for the next few millennia, focusing on size, taste, color, rind thickness, and other traits, has given us the intensely sweet, red watermelon we know today.

GENES, DISEASE, AND RISK

Let's return to humans and to traits more important than height or even sweetness. Genetics plays a major role in many maladies of current concern, like cardiovascular diseases and cancers—the two leading causes of death globally—as well as diabetes, Alzheimer's disease, and numerous mental disorders. In some cases, there are single genes that have an oversized role in susceptibility. Altered sequences in genes named BRCA1 and BRCA2 (short for breast cancer genes 1 and 2) are found in about 5% of breast cancer patients, and women with these variants have a roughly fivefold greater risk of developing breast cancer at some point in their lives compared to the general population. It's more often the case, however, that there aren't just a few genes at play. Rather, as is the case for height, a host of genetic variants each makes a small contribution. SNP-based studies have revealed a web of connections between an individual's genome and disease that promises to be further illuminated by whole genome sequencing. As with height, these connections are probabilistic: we can't say with certainty that a person will develop coronary artery disease or not, nor will we likely ever be able to, given that nongenetic factors, such as diet and exercise, are also important. However, we can acknowledge the spread in possible outcomes and speak in terms of variation for groups of people, as in our height graphs above, or risks for specific individuals.

Given one's DNA sequence, how much greater or lower is one's probability of developing type 2 diabetes, for example, compared to the average person? In 2018, a group led by Sekar Kathiresan at Massachusetts General Hospital and Harvard Medical School showed that, for a host of diseases, an SNP-based score derived from the same UK Biobank cohort noted earlier can robustly identify high risks; for example, a fivefold increase in the likelihood of developing coronary artery disease. A factor of five is comparable in magnitude to rarer single-gene variants that are already relevant to clinical screening and diagnosis, like BRCA mutations. The researchers proposed that "it is time to contemplate the inclusion of polygenic risk prediction in clinical care." They note, however, that the data so far used in their study and others are based primarily on people of European ancestry, and therefore the predictions derived from them are less accurate for members of other ethnic groups. To similarly treat a broader range of people, important for both individual health and social equity, we need more large-scale genomic studies across a wide range of ethnicities.

Of course, an elevated risk uncovered by a genetic test is just that—a risk and not a certainty, a probability connected to the development of some disorder. Its utility lies in the hope that the identification of risk will suggest changes in lifestyle or will direct the application of diagnostic or preventive tools specifically to those likely to need them. For various cancers, for example, early detection improves the odds of successful treatment, but the detection methods themselves can carry harm. Rather than indiscriminate testing on everyone, using genetic scores to identify candidates for whom the risks from disease outweigh the risks from detection would be, on average, beneficial to everyone.

PICK A GENOME

If you find yourself with a genome that puts you at high risk for various problems, you might ask yourself, Could it have been different? Your genome is a random composite of your parents' and grandparents' genomes. The die was cast long ago—too late for you to do anything about it. The genome of your future children, however, has yet to be fixed, raising the question of whether you could guide its DNA sequence. In the next chapter, we'll see how to rewrite DNA sequences to be whatever we like. Here we look at the more limited, though easier, method of selecting one particular genome from the set of possibilities. This selection is, to use a loaded term, unnatural. It's based on another unnatural procedure that caused shock and outrage upon its introduction, but that is now widely accepted and appreciated: in vitro fertilization.

In 1969, Robert Edwards, Barry Bavister, and Patrick Steptoe announced that they had successfully fertilized human egg cells with human sperm cells in vitro—meaning outside of a human body, or any other body. (*In vitro* literally means "in glass" and is the standard term for experiments performed in an artificial environment, like a petri dish or a test tube.) The second half of the two-sentence abstract of their paper drily notes that "there may be certain clinical and scientific uses" for the procedure. The obvious use is the treatment of infertility, as the authors were well aware. Steptoe, in fact, was a gynecologist who worked closely with women suffering from fertility problems. It took nearly another 10 years of work, led mainly by Edwards, Steptoe, and a nurse named Jean Purdy, until the steps of egg cell gathering, external fertilization, and implantation into the mother all fell into place. The first "test tube baby," Louise Brown, was born in 1978 to an infertile couple, making headlines around the world. The intervening decade saw a great deal of discussion, filled with both excitement and trepidation, concerning the new intersection of technology and conception. A 1969 *Life* magazine cover featured an illustrated fetus next to a photograph of a mother and child, with text posing questions about what would happen as a result of the "new methods of human reproduction," including "Will children and parents still love each other?" As if to answer, a poll of Americans in the same issue reported that only about 60% of respondents (55% of men, 61% of women) believed that a child born through in vitro fertilization "would feel love" for his or her family. (It doesn't state how many naturally conceived children the survey group believed love their families.) Only about a third approved of in vitro fertilization methods for infertile parents. By 1978, however, according to a Gallup poll also conducted in the United States, 60% had a favorable view of in vitro fertilization, and just over half reported that they would be willing to try it if they were infertile. These days mention of technologically assisted reproduction hardly raises an eyebrow. There have been in total about eight million babies born worldwide through in vitro fertilization as of 2018. In the United States, about 2% of births each year are "test tube babies"; Denmark has the highest rate, about 9%. As far as I know, no one considers the resulting people as anything less (or more) than human.

It may strike us as silly that it could have ever been otherwise, but issues of creation and reproduction tap into deeply held and often unstated notions of identity and humanity. Our views can change, however, as we examine them critically. We've been aided in this case by the harmlessness that decades of in vitro fertilization methods have demonstrated and also, I suspect, by a growing realization of the physical nature of biology. From the perspective of their structure, their function, or the information they encode, it doesn't matter where strands of DNA happen to be brought together—in vitro or in utero, in a petri dish or in a body.

As described so far, there's no connection between in vitro fertilization and knowledge of genetic traits. In practice, in vitro fertilization involves the fertilization of multiple eggs, typically around a dozen. The collection, handling, and implantation of eggs can all fail in various ways, so one works with several to minimize the odds of ending up with zero viable candidate embryos. These fertilized eggs have different genomes thanks to the shuffling of DNA in sperm and egg cells described earlier. What if we could peek at these nascent genomes? Could we avoid choosing an embryo with a high risk for a debilitating disease, or select a particular embryo with traits we want for the child? This peeking is, in fact, possible.

At three days postfertilization, each of us consisted of just six to ten cells. If one of these cells were removed, we'd be fine, even though one cell is a sizable fraction of the whole. Again, we see the wonders of self-assembly that we encountered in chapter 7—the remaining cells and their progeny fill in the gaps and pick up the cues corresponding to their locations, not their ancestry, and develop into the proper ensemble of cells characteristic of the organism. One can perform this extraction with fine pipettes under a microscope, as shown, holding the embryo with one pipette and applying suction with another to gently remove a single cell. Alternatively, one could wait a few more days, until there are around a hundred cells that have differentiated into two distinct groups: a cluster that will develop into the fetus and a shell that will eventually form the placenta. Extraction of 10 or so cells from the perimeter leaves the progenitors of the fetus completely undisturbed and still reveals the child's genome. Compared to the single-cell extraction, there's less time available between this procedure and the implantation of the embryo in the mother, so the genomic analysis must be fast. As we've seen, however, techniques for reading DNA are increasingly rapid. With either method, one gathers cells that can report on the genetic makeup of their embryonic cohort.

If there is more than one viable embryo to choose from for implantation, as is frequently the case, we could in principle use the information gathered from the embryonic biopsy and DNA characterization to choose among them, rather than relying on random chance. This is, in fact, already widely done to screen for embryos that have a high risk of single-gene-derived disorders, such as cystic fibrosis. To be concrete, imagine three embryos, one of whose genomes contains the transporter gene mutation that leads to cystic fibrosis; the other two do not. All three are equally unnatural from the perspective of being in vitro fertilized, or equally natural from the perspective of arising from the normally accessible combinations of parental genomes. There isn't any genetic editing involved; no DNA sequence is changed. There is, however, an element of choice—the deliberate selection of one genome over another.

There are some disorders that are even easier to diagnose. Genomes aren't single, continuous strands of DNA, but are divided into pieces called *chromosomes*, as noted in chapter 3. Humans have 46 (23 pairs), but errors in cell division sometimes lead to a missing or an extra chromosome. This is usually fatal—the fetus doesn't survive—but not always. Down syndrome, for example, arises from having three of chromosome number 21 rather than a pair. The extra genetic material leads to an overabundance of the proteins encoded by genes on chromosome number 21, which manifests itself in a wide range of neurological and physical symptoms. Detecting an extra or a missing chromosome is easy. In fact, it doesn't require in vitro fertilization or removing any cells from an embryo. In a normal pregnancy, one can sample the amniotic fluid surrounding the fetus, which contains more than enough sloughed-off fetal cells to assess their genetic architecture.

While peeking at embryonic genomes in vitro can, and does, allow for selection of embryos for which genetic signatures of particular traits are clear (for example, single-gene determinants or alternations in chromosome number), would it work for more complex traits? In principle, yes. In practice, the answer is more nuanced.

As we've seen, the genetic component of many disorders is determined by thousands of genes, ensembles that are being revealed by present-day mapping studies. We can identify the rare, high-risk genomes, let's say the 1 per 200 that correspond to a fivefold greater risk of contracting coronary artery disease, a level already considered actionable for single-gene factors. These extremes of risk, therefore, are within the range of present-day embryo selection, where their result is not a course of diagnosis or treatment but rather preventing occurrence entirely. Again, given three embryos, one that has a high probability of eventually contracting a disease while the other two do not, we could select one of the latter for implantation. Note, however, that in the vast majority of cases none of the three embryos would show a high risk; it's the rare occurrences that we'd occasionally find and might select against. Neither the numbers nor the method are hypothetical. The prevalence and risk factor of coronary artery disease are taken directly from the 2018 SNP-based study noted earlier.

One might think that if selecting against harmful traits is feasible, selecting for desirable traits would work just as well. This is not the case. The symmetry is broken by the nature of randomness. Let's again consider height, though the same argument holds for hair color, intelligence, or any other characteristic with a large hereditary component governed by a large number of sites on the genome. A set of three embryos samples three of the vast number of genetic rearrangements that could have occurred. It is three instances of flipping 10,000 coins, representing the roughly 10,000 genetic determinants of height. It's possible that you'd find in one of the three instances 10,000 heads, the exceptional child a foot taller than average, but it is astronomically unlikely. It's vastly more probable that you'd find 4987 or 4672 or 5115 heads. It's true that among these we could select the one with the most height-promoting variants, the 5115-head case in our coin analogy, but the net effect would be a tiny fraction of the coordinated pull of all 10,000 variants, and would, moreover, likely be swamped by the nongenetic variability also at play. Screening to reject an extreme genome in the unlikely event that it turns up is very different than hoping that a few random genomes will happen to be extreme.

More precise analyses reach similar conclusions. A 2019 study led by Shai Carmi at the Hebrew University in Jerusalem found that, with current technologies and data, one might expect embryo selection to give a height advantage of about 1 inch and an IQ advantage of 2.5 points, overall a "limited utility."

THE IMPLICATIONS OF CHOICE

Influencing the traits that will make up a future child is both awe-inspiring and frightening, and it's not surprising that the notion of "designer babies" has captivated the public. There are deep ethical issues that intersect embryo selection, a full discussion of which could fill multiple books. My goal here is to describe the science that should underlie each of our assessments of how to apply, or not to apply, new biotechnologies, and not to fully survey the field of biological ethics. Still, I'd be remiss if I didn't at least outline some of the key concerns.

The notion of shaping future people has a long and dispiriting history, most notoriously manifested in the genocidal philosophy of the Nazis but also applied in nominally free democracies in the early twentieth century. People deemed "weak-minded," for example, could be forcibly sterilized in several US states, a procedure declared legal nationwide by the Supreme Court in 1927. About 30,000 forced sterilizations had been performed in the United States by 1939. Several other countries had similar policies. The promoters of the eugenics movement were optimistic that these activities would broadly benefit society and aid human well-being. Supreme Court Justice Oliver Wendell Holmes wrote, "We have seen more than once that the public welfare may call upon the best citizens for their lives. It would be strange if it could not call upon those who already sap the strength of the State for these lesser sacrifices, often not felt to be such by those concerned. . . . The principle that sustains compulsory vaccination is broad enough to cover cutting the Fallopian tubes." Whatever the justifications its proponents invoked, however, the implementation of forced sterilization tended to target the poor and others on the margins of society, conveniently applying flimsy or capricious assessments of moral or intellectual character to quash the rights of those deemed unworthy. The Nazis stretched notions of genetic culling to even more horrific extremes. These historical examples should inform our present-day decisions, because of their similarities but also because of their differences.

One key difference is that, at least so far, modern tools of embryo selection are driven by individuals rather than the state, with prospective parents making decisions about their reproduction, subject to legal constraints. This brings up, however, issues of access—whether embryo selection technologies will disproportionately serve the wealthy. This is a concern with all health technologies, including in vitro fertilization, which can be ameliorated by decreasing costs and ensuring widespread medical coverage. Embryo selection also intersects more general issues regarding the relationships between parents, children, and society. Most people consider it acceptable, for example, for parents to buy their children athletic training, music lessons, or tutoring. Whether nudges obtained before birth are similar or categorically different from other such benefits that better-resourced parents can confer on their children isn't obvious and calls for informed deliberation.

Even if embryo selection is an individual choice, it can alter the overall makeup of the human population simply by diminishing the propagation of genetic variants to future generations. Of course, we already influence the makeup of the gene pool by choosing our partners, and the signatures of "assortative mating," in which people of similar educational or socioeconomic backgrounds pair up, are evident in genetic analyses of populations. In a sense, embryo selection makes this more deliberate. If, extrapolating wildly, no children at all carrying the BRCA1 or BRCA2 mutations that lead to high breast cancer risks are born, these gene variants won't be part of the pool that makes up subsequent generations. The potential benefits are clear, but there may be downsides as well. It's conceivable, though unlikely, that these variants could be beneficial in some unforeseen context, perhaps some future plague against which the aberrant BRCA1 or BRCA2 gives resistance. More plausibly, the absence of the traits we select against may be regrettable in itself. No one will mind if breast cancer disappears from the earth, but what about Down syndrome, which despite its difficulties is compatible with a joyful and fulfilling life? Many works of brilliance and beauty have been created by people who suffered from depression, a trait that is partly hereditary. We may lose some of the richness of humanity by eliminating disorders like this, reducing genetic diversity in favor of some narrow conception of normalcy. How "normal" is defined by society, and who gets to decide, further complicates the issue. On the other hand, no one would argue that we should deliberately increase the number of people affected by Down syndrome or depression, leading to the question of why the status quo is optimal. Finally, even if we all agree that preserving human genetic variation is important, the notion that reproductive autonomy is subservient to the public good brings back echoes of last century's eugenics. These are exceptionally difficult issues to think about. I encourage the reader not to look away from them, unsettling as they may be, as their already sizable relevance to our society will only grow in the future.

The individualistic nature of embryo selection highlights the importance of education. As we've seen, concepts of variability, uncertainty, randomness, and the very nature of genes and genetic traits are central to knowing what can and can't be done with modern tools. Misunderstanding any of these can lead to false hopes and misplaced expectations that may end up on the shoulders of future generations. Education is crucial. This is perhaps the least controversial statement in all of biological ethics, but how and when to implement education isn't obvious. My own belief is that an understanding of cells, genes, development, and technology should be part of the tool kit carried by every educated adult and not merely the subject of discussion at a fertility clinic.

All the concerns outlined above are important, interesting, and challenging. None, however, involve genetic engineering; none involve changes to DNA itself. The mere knowledge of the possible and the actual physical rearrangements of genomes, coupled with tools to handle cells and read DNA, gives us the means to reshape life in ways that would have been inconceivable a generation ago. The ability to rewrite DNA opens further possibilities. In the next chapter, we see how this is done.

How We Write DNA

We routinely reshape organisms. We reset broken bones, tie saplings to posts, and feed antibiotics to livestock to make them grow larger. These actions affect proteins, cells, and tissues and influence whether genes are turned on or off, but they don't change the sequence of As, Cs, Gs, and Ts that make up the organism's genome. Now, however, we have the tools to make such alterations and rewrite DNA, revising the instruction set for life's components directly.

Ever since the first chapter, we've seen that, for DNA and all the other materials that make up living things, biological function and physical form are inseparable. When examining the astounding tools with which to read DNA sequences, I noted that their invention required taking seriously the notion of DNA as a physical object, developing techniques to cut, grow, move, and monitor it that take into account its material characteristics as well as overarching themes related to self-assembly and randomness. This perspective applies to gene editing as well, as we seek to modify a strand of DNA. Here, however, we manipulate this molecule inside living cells, rather than in a machine, calling for a different set of approaches. In this chapter, we see how to edit DNA using a revolutionary twenty-first-century technology known as *CRISPR/Cas9*. To understand why this tool is so stunning, we first look at the methods that preceded it, remarkable in themselves.

HIJACKING BACTERIA

The heroes of the gene editing saga are bacteria, whose abilities have amazed us several times already in past chapters. There are many species of bacteria that are easy to grow in vast quantities, such as the lab stalwart *E. coli.* A warm bucket of nutrient-rich broth can sustain trillions of *E. coli*, each growing, dividing, and making lots of proteins. The proteins are those encoded in its genome—proteins that digest sugars, propel the cells through liquid, build membranes, and more. What if we could give *E. coli* a human gene, with which it would create a human protein, such as insulin? The bacterium would become a small, living, nearly infinitely replicable pharmaceutical factory.

The insulin example isn't hypothetical. People with type 1 diabetes don't produce insulin, leading to a debilitating and potentially fatal inability to regulate blood sugar levels. Since the 1920s, diabetics had been treated with insulin from pigs and cattle, the purification of which is difficult and expensive. In the 1970s, even after decades of technical refinements, preparing one pound of insulin required piles of pancreases from more than 20,000 animals. Moreover, the nonhuman material often induced allergic reactions in patients. Like the sonic hedgehog proteins we saw in chapter 7, the amino acid sequences of the animal insulins are very similar to the human insulin sequence—similar enough to work—but they aren't exactly the same, and these slight differences along with other substances carried by the animal extractions can trigger the immune system to mount a vigorous response. For all these reasons, the idea of mass-producing human insulin was tantalizing. For all of our species' history until the 1970s, however, humans were the only large-scale source of human proteins. That changed when we learned how to move genes between species, starting with the transformation of bacteria.

I'll describe in some detail the process of turning bacteria into machines that serve our ends, because the philosophy behind it is so different from our approach to nonbiological engineering. In a *Calvin and Hobbes* comic strip, the six-year-old Calvin asks his father, "How do they know the load limit on bridges, Dad?" The reply: "They drive bigger and bigger trucks over the bridge until it breaks. Then they weigh the last truck and rebuild the bridge." Calvin is stunned, and we're amused, since this is ridiculous. Similarly, no one would build a car by randomly connecting engines, axles, wheels, and other components in different arrangements and searching for the one-in-a-million that turned out to be a functioning vehicle. This is, however, a sensible approach to bioengineering, and it's made possible by the ability of living materials to self-assemble and by the modular, reproducible components with which they build themselves.

Bacteria possess powerful tools for manipulating DNA, including proteins called *restriction enzymes* that cut the double helix. The cutting isn't haphazard; each restriction enzyme recognizes a specific DNA sequence, typically a sequence present in the genome of a virus that can infect the bacterium. The restriction enzymes, therefore, serve as a defense against viral invasion, part of an age-old struggle to which we return later in the chapter. Most restriction enzymes leave the cut ends of the DNA "sticky"—that is, in fact, the technical term—ready to adhere to other cut ends. The stickiness isn't from some artificial adhesive but is a consequence of the shape of the cut: one of the single strands of the double-stranded DNA overhangs the other. The overhanging sequence, therefore, can bind to another cut DNA strand, as long as the two overhangs are complementary—As matching Ts and Cs matching Gs.

Another idiosyncrasy of bacteria is a fondness for taking up DNA from their surroundings, potentially scavenging useful traits from now deceased neighbors. Loops of DNA much smaller than the full genome are particularly amenable to being transported in or out of bacterial cells. Many bacteria contain such loops, called *plasmids*.

This repertoire of tools lets us move a human gene into a bacterium. First, one acquires the segment of DNA corresponding to the gene of interest—human insulin, for example. This can be cut out of an existing genome or synthesized from scratch—tractable even in the 1970s for a small gene like insulin. Recall that, thanks to the polymerase chain reaction, another technology enabled by microorganisms (chapter 1), even a tiny amount of DNA can be duplicated to give millions of identical copies. Restriction enzyme cleavage sites flanking the gene sequence of interest provide a multitude of sticky DNA fragments. Meanwhile, the researcher also collects plasmids from bacteria, grown in large numbers, and similarly make cuts in these. The vials are mixed together, and the target gene fragments and the opened plasmids meander through their watery surroundings, encountering each other through the randomness of Brownian motion. Some stick to each other to give new loops, composed of the original DNA plus the target gene (the latter indicated by curved bars in the illustration).

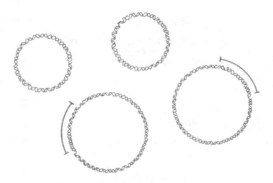

Some loops may close without grabbing the target gene fragment; these duds won't matter, as we'll see shortly. At least some of the loops will have incorporated the sequence we want. (If you look carefully, you'll find gaps drawn in the DNA backbone where the sticky ends meet; bacterial proteins repair these gaps.)

The next task is to get bacteria to swallow the plasmids. Uptake is rare, but can be enhanced by pulses of heat or electricity, both of which open transient pores in the bacterial membrane. Still, only a tiny fraction of the bacteria will find themselves in possession of an engineered circle of DNA. One can, however, winnow the crop to keep only these microbes, killing all the rest. The researcher selects or designs the plasmid to have other genes present, for example, genes for proteins that confer resistance to antibiotics. Exposing the bacteria to antibiotics, therefore, leaves alive only those that have taken up the plasmid. It's fine if these are few in number; bathed in broth, billions more grow. Of course, some of these surviving bacteria will have plasmids that closed without capturing our target gene, but straightforward techniques for determining the size or the sequence of the loop let the researcher identify and reject these cases, and again grow billions of the correct bacteria.

At the end of this process of cutting, gluing, and sifting, one is left with an organism that didn't exist before: a bacterium encapsulating and expressing a foreign gene. Working together, the labs of Stanley Cohen at Stanford University and Herbert Boyer at the University of California at San Francisco announced the first successful bacterial transformation in 1973. The potential soon became clear: microbes could be turned into microscopic assembly lines, churning out materials they wouldn't naturally make. Because insulin was an especially appealing target, several labs raced to genetically engineer microbes to produce it. Walter Gilbert's group at Harvard University—the same Walter Gilbert who pioneered DNA sequencing (chapter 13)—used DNA purified from humans to provide the insulin gene, an approach subject to stringent regulations on its handling. A general wariness about moving genes between species led to a moratorium on all such work in Cambridge, Massachusetts, the town in which Harvard is situated, following lively city council meetings. As one of the councillors said later, "I tried to understand the science, but I decided I couldn't make a legitimate assessment of the risk. When I realized I couldn't decide to vote for or against a moratorium on scientific grounds, I shifted to the political." He voted for the moratorium. Meanwhile, in a makeshift lab a few miles south of San Francisco, a small biotech start-up used chemically synthesized DNA for the insulin gene, avoiding the regulatory and social drama associated with human-derived molecules. The arrangements of atoms into As, Cs, Gs, and Ts are exactly identical, regardless of what process made them—a molecule is a molecule is a molecule—but law and public opinion don't necessarily recognize this. The company, Genentech, was founded by Boyer and venture capitalist Robert Swanson, and in August of 1978 its vats of microbes created the first-ever bacterially produced human insulin proteins. The small firm partnered with pharmaceutical giant Eli Lilly to tackle clinical tests, manufacturing, and administrative approval. In 1983, its insulin was brought to market, revolutionizing the treatment of diabetes and becoming the first human therapeutic drug produced by genetic engineering.

The bacterial production of insulin opened the floodgates to biologically produced pharmaceuticals, designed to treat an ever-expanding range of diseases and making up a market currently worth around $250 billion. These days insulin itself is mostly produced from engineered yeast rather than bacteria, as are many other drugs. Yeast are also single-celled microorganisms but are eukaryotes like us. We share with yeast various machineries for modifying proteins that bacteria lack. Insulin is initially formed in your pancreas as a single amino acid chain (chapter 2) that is subsequently cut into three pieces, two of them linked by new chemical bonds and one discarded. Bacteria can't perform these chemical steps, so the Genentech researchers engineered the two final fragments as separate genes whose resulting proteins joined themselves together. Yeast allow a more direct and productive approach.

SPLICING NICER RICE

It took further insights and inventions to enable the insertion of genes into eukaryotes. Few eukaryotes have plasmids, and eukaryotic genes generally need to be inserted into chromosomes, becoming part of the overall genome, for cells to express them. (This insertion can be relevant to bacterial engineering as well: for really robust modification of bacteria, one wants the gene of interest to be integrated into the main bacterial genome, not into plasmids that might be lost as cells divide.) We've had methods for modifying eukaryotic genomes for the past few decades, but until recently they've been inefficient, imprecise, and labor intensive. CRISPR/Cas9 has changed all this, so I won't say much about other methods except to note the general tactics and to give an example of why one might go to the trouble of using them.

If we think of a eukaryotic genome as a large library whose stacks are closed to us, like the Library of Congress or a private collection, our task is to somehow sneak a new book of our own onto the shelves. We could try leaving our book in a public part of the building, but it's unlikely that a librarian would decide to shelve it; eukaryotic cells don't take up and integrate random scraps of DNA. We're more likely to succeed if the library is renovating or relocating, and amid the shuffle our book is grouped with the others. Newly fertilized egg cells, in the interval before the parents' DNA comes together, provide such an opportunity. DNA carefully injected into the nascent embryo can find itself inserted into the genome. When first developed, this technique had a low success rate and allowed only insertion at random locations in the genome, potentially disrupting existing genes; but refinements have made it more robust and have even enabled some degree of targeting. The approach is a standard one for generating transgenic mice, for example, with fluorescent protein genes that serve as reporters of cellular activities (chapter 2).

Another approach is to recruit someone else who can more easily enter the library—perhaps a skilled burglar. We often employ viruses. Viruses infiltrate cells and replicate their genomes, either as stand-alone pieces or integrated into their host's genome. Being small and efficient, viruses don't easily lend themselves to the addition of new cargo, but it can be done, with the modified viruses delivering genes to cells in a variety of different organisms. Viral insertion of DNA has advantages over microinjection in that the virus itself, rather than a lab technician with a fine needle, takes care of getting material into the cell, and the cell type needn't be restricted to newly fertilized eggs. It suffers similar disadvantages, though, of not being as reliable as we'd like and being quite random regarding where in the genome the delivered gene is inserted. If we're making transgenic mice, we don't mind if our success rate is far from perfect. Again we winnow, studying only those in which the transformation has worked. If we want to create a human therapy, however, our standards for robustness and precision are much higher.

Bacteria, even those that can enter other cells and cause harm, can't in general alter eukaryotic genomes. There are rare exceptions, however. The soil microbe *Agrobacterium tumefaciens*, when the opportunity arises, infects plants. The microorganism snips a certain piece of its own DNA and injects it into a plant cell, along with proteins that direct the package to the nucleus and induce the plant to use its DNA repair machinery to integrate the bacterial DNA into its genome. The eventual consequence is the development of tumors that encourage bacterial growth. Scientists have engineered *Agrobacterium tumefaciens* in which the tumor-causing genes are replaced by whatever genes we like, turning the bacterium into a potent, and harmless, gene delivery tool for plants.

One of the most interesting applications of *Agrobacterium*-mediated gene delivery is the engineering of rice to fight vitamin A deficiency. Inadequate vitamin A causes blindness in 250,000 to 500,000 children annually—it's the largest cause of preventable blindness in children. About half of these kids die within a year of becoming blind, tragically highlighting the overall importance of the vitamin to health. Our bodies produce vitamin A from various precursors, most notably beta-carotene, which contributes to the orange color of foods like carrots and sweet potatoes. Beta-carotene is not, however, found in rice, a cheap, abundant staple food in many regions where vitamin A deficiency is prevalent. Rice plants are, in fact, naturally capable of making beta-carotene—they do so in leaves, where the molecule plays a role in photosynthesis. They don't, however, express the relevant genes in the starchy grains we eat. Researchers led by Ingo Potrykus of the Swiss Federal Institute of Technology and Peter Beyer of the University of Freiburg, Germany, therefore developed "golden rice," adding to the rice genome two genes and their promoters, derived from the daffodil and a bacterium, that lead to beta-carotene synthesis in the edible part of rice. After years of work, the success of the project was announced in 2000. Further improvements developed in conjunction with the biotechnology company Syngenta increased the beta-carotene levels by over a factor of 20; Syngenta subsequently donated all patents, technologies, and transgenic seeds associated with golden rice to the public.

Clinical studies showed that the strikingly yellow-orange rice is safe and effective, delivering beta-carotene that the human body converts into vitamin A. The American Society for Nutrition noted that "Golden Rice could probably supply 50% of the Recommended Dietary Allowance (RDA) of vitamin A from a very modest amount—perhaps a cup—

of rice" per day. Despite this, adoption of golden rice has been exceptionally slow; the plants are anathemas to various groups opposed to genetically modified organisms. Especially in recent years, the complaints are not about the plant per se, whose utility is hard to deny, but rather that its use opens the door to other modified foods, or that it would be better if vitamin A were supplied by providing the poor with a generally more nutrient-rich diet. Still, there is some motion. Bangladesh, 21% of whose children suffer from vitamin A deficiency, became in 2019 the first country to approve the planting of golden rice seeds. In the Philippines, the fraction of vitamin A–deficient children between six months and five years of age increased from 15% to 20% between 2008 and 2013, numbers that correspond to blindness and death for thousands of children. In 2019, the Philippines acknowledged the safety of golden rice, setting the stage for its planting. Twenty years after the invention of golden rice, debates and controversies surrounding it and other genetically modified crops continue, informed in some cases by an understanding of the underlying science, in some cases by complex issues of economics and commerce, and in many cases by amorphous perceptions and opinions. More recent technologies offer even more scope for discussion.

CRISPR AND THE GENE EDITING REVOLUTION

As we've seen, it's been possible for decades to alter genomes in all sorts of organisms. The methods I've sketched are difficult and inelegant, though. It takes trial and error to get a creature to take up a new gene, and where in its genome the gene ends up is either random, with random consequences for its expression, or crudely controlled at the cost of even more tedious engineering. For a bacterium or a mouse into which we'd like to insert an insulin gene or a fluorescent reporter, it is perhaps acceptable to keep trying until we get the outcome we want, but this strategy has its limitations, especially if we dream of curing genetic diseases in humans. What we'd really like is a way to edit genomes simply and precisely, identifying specific stretches of DNA and either replacing them with our own designs or cleanly excising them.

This ability is now in our grasp. I focus on the CRISPR/Cas9 system, but as is often the case with technology, more than one revolutionary method emerged at around the same time. Here the alternatives, *zinc finger nucleases* and *transcription activator-like effector nucleases* (TALENs), aren't quite as fast, cheap, and easy to use as CRISPR/Cas9; I mention them both for completeness and to suggest that the atmosphere of the early twenty-first century was full of the seeds of genome editing, ready to sprout.

CRISPR/Cas9 is an ancient technique, or a very new one. It was discovered, or it was invented. Each of these alternatives is correct depending on one's perspective. We first meet the primordial practitioners of genome manipulation before turning to their modern, human-driven manifestations.

Nature, to borrow Tennyson's phrase, was "red in tooth and claw" long before teeth and claws existed. Even microorganisms battle one another. Bacteria stab and poison their fellows, amoebas consume bacteria, viruses infect cells and vandalize their genomes. In addition to weapons, all these creatures have developed defenses. Until recently, it was thought that bacterial defenses act solely in the present, detecting cues as they occur without any memory of past insults. We now know, however, that bacteria do remember. Like our own immune systems, they keep a record of prior antagonists to enable rapid responses if they're encountered again. The discovery of this bacterial immune system is in itself emblematic of the modern biotechnological age: it was made possible by DNA sequencing and computers.

In 1987, a Japanese team studying a gene in the workhorse bacterium *E. coli* noted in passing "an unusual structure," a sequence of 29 nucleotides that repeated five times, with each repetition separated by 32 nucleotides. What its purpose might be, and whether it exists in other genomes, were mysteries. The observation attracted little attention; life is full of oddities, mostly of no consequence.

In the 1990s, Francisco Mojica and colleagues in Spain discovered similar repeating DNA patterns in the genomes of several archaea. Archaea are single-celled creatures without nuclei or organelles (chapter 5). While superficially resembling bacteria, they form a distinct branch of the tree of life; like us, they are separated from bacteria by a few billion years of evolution. Mojica came across the Japanese group's paper and realized that the presence of a similar genetic structure in such different organisms was a clue that the pattern plays an important role in the lives of microbes. In the late 1990s, as increasing numbers of bacterial genomes were sequenced and computational tools allowed one to pore over them at a desktop computer, Mojica and colleagues discovered many more instances of spaced, repeating DNA in bacteria and archaea. Their persistence and the puzzle they uncovered didn't deliver fame and fortune, however; Mojica's lab struggled for years to get funding, with the lack of relevance of the repeating sequences being a persistent critique. The genomes continued to pile up, however. In 2002, Mojica and Ruud Jansen of Utrecht University coined the term *CRISPR* for the genomic entities: clustered regularly interspaced short palindromic repeats. Researchers also noticed that a handful of genes were often found in the neighborhood of the spaced repetitions; these were named *Cas* (CRISPR-associated) genes. Though CRISPR now had a name, its function was still a mystery.

Again, insights came from genomes and computers. Three papers published in 2005, one from Mojica's group, one from Christine Pourcel and colleagues in France, and one from Alexander Bolotin and colleagues, also in France, announced that the DNA sequences of spacers matched nonbacterial and nonarchaeal DNA sequences, especially those of viruses. Since time immemorial, viruses have infected cells, and the viral signature suggested that CRISPR may be part of some sort of previously unrealized defense strategy. Though attention would slowly build, none of these papers were particularly well appreciated at their advent. Mojica's, the first to be written, was repeatedly rejected from prestigious journals before eventually finding a more modest home. Nonetheless, the essence of CRISPR was beginning to be unraveled. Why keep a carefully curated arrangement of pieces of DNA from viruses that infect you? So that you can use these mug shots to recognize and deactivate the viruses, should they invade again.

As a bacterial immune system, CRISPR/Cas works by making use of the complementarity of DNA and RNA (chapter 3), as well as protein machines that can cleave DNA. First, fragments of DNA encountered within the cell—likely from viral intruders and not the bacteria's or archaea's own DNA—are cut and inserted into the genome in between the repeating units by the proteins Cas1 and Cas2. The spacers, therefore, delineate a library of past viral encounters. Next, the cells take the seemingly foolhardy step of transcribing sections of the repeats and the spacers into RNA, called crRNA (for CRISPR RNA). Recall that transcription (chapter 3) is typically the first step to making proteins— potentially a dangerous activity if viral genes are the starting point. The RNA of the repeat segment, however, is in part complementary to RNA transcribed from another region in the CRISPR neighborhood of the bacterial or archaeal genome, called tracrRNA. The crRNA and tracrRNA together form an RNA double helix, with the viral-derived segment dangling unbound (left in the illustration). The CRISPR-associated protein, Cas9, recognizes and binds this RNA assembly (middle).

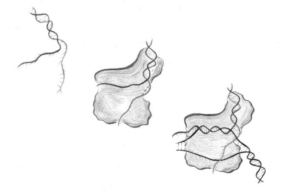

The RNA-Cas9 complex meanders around the cell thanks to Brownian motion. If it encounters a specific, short, and fairly common DNA sequence among the molecules it collides with, it attaches; this same DNA motif formed part of Cas1 and Cas2's identification of where to excise DNA. Cas9 destabilizes the bonds between the two strands of the DNA it has latched onto, essentially melting the double helix (chapter 1) in a small neighborhood. If the single-stranded RNA that Cas9 has been carrying is complementary to the opened-up DNA, which is only the case if the DNA is of the same virus from which the crRNA was derived, the two strands form a DNA-RNA hybrid. The peculiar but stable double-helical construction lies nestled in a groove in Cas9, lined with positive electrical charges that hold the highly negative DNA and RNA in place (right; DNA in dark blue).

Cas9 also contains two sections capable of slicing the DNA backbone. As the protein grips the DNA-RNA helix, these cleavers change their orientation, each situating themselves in contact with one of the viral DNA strands, the one that's joined with the RNA and the one left over, its former partner. I've illustrated the most dramatic shape change, in which one of the cleavers, depicted in green, rotates nearly 180 degrees. Cas9 makes the cuts and the viral DNA is crippled, with whatever genes the segment contains rendered nonfunctional. Cas9 wanders off, its job complete.

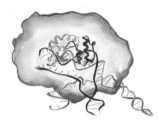

There's variety among CRISPR sequences and Cas proteins that I won't go into. The basic mechanism is the same for all the variants, and I focus on the relatively simple and elegant Cas9 protein, which has become the workhorse of most biotechnological applications.

As a bacterial immune system, CRISPR/Cas9 is impressive, combining memory and defense. At a deeper level, it reveals that Nature, billions of years ago, solved one of the grand challenges of genetic engineering: finding one particular DNA sequence among billions of possibilities, and then altering it. Here the finding involves viral DNA libraries, and the alteration is simply destruction by cleavage. Several researchers realized, however, that Nature's solutions could be tweaked to be far more generally applicable, and even constructive.

In 2012, the teams of Jennifer Doudna at the University of California at Berkeley and Emmanuelle Charpentier at the University of Umeå in Sweden, working together, published an enormously influential paper elegantly demonstrating CRISPR/Cas9's power as a DNA editor outside of bacteria or archaea. Charpentier's research group discovered tracrRNA and had deciphered many aspects of Cas9's activity. Doudna and coworkers were experts in RNA, and their wide range of research targets included the complex landscape of CRISPR-associated proteins. In their natural context, tracrRNA and virus-derived crRNA together direct Cas9. Doudna and Charpentier realized that replacing the viral sequence with whatever sequence you want programs a new destination into Cas9, and furthermore that the tracrRNA and crRNA can be replaced by a single, appropriately designed piece of RNA that folds back on itself, referred to as a single-guide RNA. Therefore, using just one easy-to-synthesize RNA strand and one easy-to-produce protein, Cas9, should enable simple, general targeting of specific DNA sequences. To cut DNA at the location of a particular nucleotide sequence, the recipe is as follows: make single-guide RNA molecules that contain that sequence (with the Ts replaced by Us, as is always the case for RNA (chapter 3)), add Cas9, mix into your system of interest, and you're done. The two research groups showed that this works in a test tube, outside the familiar confines of a bacterial cell. Reporting the results in a 2012 paper, the authors noted that the system is "efficient, versatile, and programmable," adding that it "could offer considerable potential for gene-targeting and genome-editing applications."

Doudna and Charpentier's 2012 paper, submitted to the prestigious journal *Science* on June 8 and published online on June 28, elicited a torrent of attention; it was clear that it demonstrated a breakthrough with far-reaching potential. The two researchers each received a $3 million Breakthrough Prize in Life Sciences in 2015, an award funded by Facebook's Mark Zuckerberg, Google's Sergey Brin, and other tech titans, awarded at a celebrity-laden Silicon Valley gala.

Returning to 2012, another research group, led by Virginijus Šikšnys at Vilnius University, Lithuania, submitted a paper on April 6 also describing RNA-guided DNA cleavage in a test tube by CRISPR/Cas9, again with a general version of the tracrRNA/crRNA template, though using two RNA fragments rather than a single-guide RNA, and similarly noting that the findings "pave the way for engineering of universal programmable RNA-guided DNA" manipulation. The manuscript was brusquely rejected; the authors sent it to a different journal, where it was published in late September, three months after Doudna and Charpentier's paper. In science, the acclaim often goes to those first across the finish line, and Šikšnys has been referred to as "the forgotten man of CRISPR." However, in 2018, he shared with Doudna and Charpentier the $1 million Kavli Prize for Nanoscience. It was certain that *someone* would win a Nobel Prize for CRISPR; the buzzing questions were "when?" and "who?" These were answered in 2020, as the Nobel Prize in Chemistry went to Doudna and Charpentier.

After the pioneering papers of 2012, many more appeared, from many research groups. In 2013, for example, the labs of Feng Zhang at the Broad Institute of MIT and Harvard University and George Church at Harvard Medical School demonstrated CRISPR/Cas9 in mouse- and human-derived cells. Popular accounts mushroomed as well; by 2019, there had been over 100 articles mentioning CRISPR in the *New York Times* alone. There is much more to the history of the study of CRISPR/Cas9, including patent fights and dramas that intersect broader issues of how the modern scientific enterprise works. My goal is to focus on the science and its uses, and so to the science we return.

CRISPR/Cas9 is purely destructive in our depiction so far, slicing DNA at the desired spot. This destruction can be used to deactivate a gene. Cells contain mechanisms to sense broken DNA and repair it, necessary for dealing with ever-present damage from chemical by-products of metabolism or physical dangers such as ultraviolet light. Repair is performed by proteins that join the ends of DNA fragments. The joining is error-prone, however, altering the sequence at the seam or adding or deleting nucleotides. These mistakes likely lead to a nonfunctional protein.

Repair can more constructively enable the insertion of new genes into the genome. We can load the cell with fragments of DNA that encode whatever gene we want. Repair proteins aren't fussy—they'll glue together any ends they find—so in some cases one of our fragments is inserted at the break, joined to the genome at each end. You might be thinking that "some cases" is masking a lot of uncertainty, and you're right: the probabilistic nature of the microscopic world manifests itself in many ways. Cut by Cas9, the free ends wander by Brownian motion. The proteins that sense and repair DNA damage wander as well. Whether they find a free end of the genome and a free end of the introduced DNA before finding two free ends of the genome is up to the vagaries of their random walks, though we can bias the outcome by adding lots of copies of our new gene to make it more likely to find one. As all this is happening, Cas9 is still present—it may again find its target DNA sequence and again cut it, repeating the process over and over. There's some probability of getting the final result we want, but it's by no means a certainty. As with the older methods, we need to screen lots of failures to find successes. Still, if it works, the method puts the new gene exactly where we want it in the genome.

Though barely a decade has elapsed since the advent of programmable CRISPR/Cas9 systems, researchers have already invented much better gene insertion schemes. A powerful method called *prime editing*, developed in 2019 by the group of David Liu at the Broad Institute of MIT and Harvard University, paints with a broad palette of cellular machineries. I won't go into the details, but the essence is to fuse Cas9

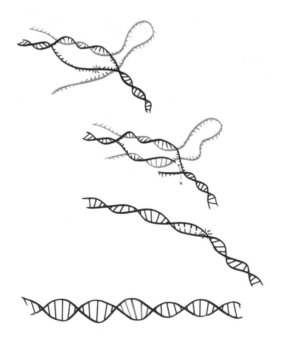

and reverse transcriptase—the protein we met in chapter 13 that makes DNA from RNA—and include as part of the guide RNA the desired insertion sequence. The chimera of Cas9 and reverse transcriptase recognizes and cuts DNA and generates new DNA at the cut. The illustration gives a flavor of the multistep process; DNA is black, RNA is gray, the insertion sequence is blue; the protein is omitted.

Liu's group demonstrated prime editing in human- and mouse-derived cells and noted that 89% of the genetic variants that are associated with human diseases are the types that can be targeted by the technique. (The remaining 11% include diseases associated with multiple copies of genes, for example, that simple rewriting won't fix.)

This elegant tailoring of DNA, making use of the needles and threads of proteins and nucleotides, is a stunning manifestation of the physicality of biological molecules. DNA is a code carrying genetic information, and to alter this code, we develop tools to quite viscerally grab, cut, paste, and glue. The precision is remarkable even from a nonbiological perspective. State-of-the-art integrated circuit chips in your computer, for example, have features as fine as about 10 nanometers. A single DNA nucleotide is about a third of a nanometer in length, and can be controllably altered by prime editing and other present-day methods.

As we've learned, simply having or not having a gene isn't the sole determinant of cellular activity. Regulation is crucial—controlling whether a gene's nucleotide code is actually read to give rise to a protein (chapter 4). Already, researchers have figured out how to use CRISPR/Cas9 to program regulation. Here again, its guide RNA serves as a navigator to specific sites in the genome. Rather than a standard Cas9, however, one uses modified variants that can't cut DNA. To turn genes off, the deactivated Cas9 simply sits on the DNA, blocking transcription by RNA polymerase like a normal repressor. The switch I described in chapter 9 that stops bacteria from swimming uses this technique. One can also turn genes on, for example, by linking a deactivated Cas9 to activators that recruit RNA polymerase. Thanks to CRISPR, we have exquisite access not only to genomes but to genetic circuits.

ON TOMATOES, T-CELLS, AND THERAPIES

Plants, the targets of genetic manipulations for millennia, can be more easily and elegantly transformed by CRISPR-based methods than by earlier techniques. The ancestor of the tomato, still existent in the wild, produces small, pea-sized fruits. Its domestication gives the larger and more nutritious tomato with which we're familiar. Selective breeding has picked particular genetic variants that we like, but at the cost of genetic diversity and tolerance to stress relative to the wild cousin. In 2018, researchers used CRISPR/Cas9 to insert just four of the genes from domesticated tomatoes into the wild plant's genome, generating fruit with three times the normal weight. The precision of CRISPR not only preserves the target genome but avoids the linkage-induced dragging along of potentially unwanted genes that is a by-product of conventional breeding, as we saw for watermelons.

The organism we care most about, of course, is us. With amazing speed, therapies and treatments based on gene editing are being developed and deployed. In addition to being useful, these techniques highlight some of the subtleties of using gene editing in practice. For all the cutting and pasting of CRISPR/Cas9 to actually happen, these molecular machines need to get inside the cells we want to modify. The most effective approach is to use viruses that are engineered to carry DNA encoding Cas9 and the requisite RNA sequences. The viruses mindlessly recognize, latch onto, and enter their target cells.

Exposing only certain cells to the viruses can be very challenging, however. One approach is to take the cells of interest out of the body, edit them, and put them back in. In 2015, researchers at the University of Pennsylvania piped blood cells and circulating immune cells out of and into the bodies of patients infected with the human immunodeficiency virus (HIV), which attacks T-cells. Years earlier, scientists noticed that a mutation in a gene called CCR5, which encodes a T-cell membrane protein, can confer protection against HIV. The Pennsylvania group extracted T-cells, disrupted the CCR5 gene, and reintroduced the modified immune cells to the patients' bloodstream. Improvements in health were minimal, but the method showed itself to be safe, spurring several groups to continue to develop the treatment. Another pioneering application in 2015 also involved immune cells, in this case removed from a donor, edited to withstand anticancer drugs, and placed into a one-year-old girl who suffered from leukemia and was unresponsive to all other treatments. Typically, donated immune cells trigger strong and potentially lethal responses in the recipient, unless perfectly matched. Here the editing also disabled the genes that drive such immune responses. The child's body adopted the modified cells, and her leukemia went into remission.

Neither of these gene editing examples used CRISPR/Cas9, but rather the slightly earlier-developed tools noted at the start of the chapter, zinc finger nucleases (for HIV) and TALENs (for leukemia). Now a flurry of CRISPR/Cas9–based therapeutics are being developed. Concurrently, treatments are moving into the human body itself, accessing tissues and organs that aren't amenable to out-of-body experiences.

In early 2020, a person with a rare genetic mutation that causes blindness became the first human to receive direct delivery of CRISPR/Cas9, via viruses injected into the eye. The mutation is in a gene that encodes a protein necessary to construct the towers containing light-sensitive proteins in certain cells of the retina. To ensure that gene editing occurs in the appropriate cells, the Cas9-encoding DNA includes a promoter sequence specific to transcription factors expressed in those retinal cells alone. If the clinical trial, yet to conclude, is successful, it will mark a milestone in our ability to treat blindness, and one achieved not through complex surgeries or electrical implants but by reconfiguring a person's own instructions for self-assembly.

CRISPR AND ANTI-CRISPR

Understanding the role of the biophysical principles of self-assembly and regulation gives us insight into how gene editing works. I've noted in passing another theme, that of randomness, in the context of Cas9 finding specific stretches of DNA. Randomness plays a larger role, however, and accounting for or controlling it is a challenge for gene editing applications. I've said that Cas9, via the guide RNA it cradles, binds to a specific DNA sequence complementary to the guide sequence. This is true, but it is an oversimplification. Like all molecular binding, the affinity is greatest for a perfect match but isn't exactly zero for an imperfect match. There is always some probability for Cas9 to target the wrong DNA. Whether this probability matters or not depends on its magnitude and on timescales. Suppose the chance of off-target Cas9 binding is 1 in 1000, and for whatever our goal may be, 1 in 1000 is an acceptable error rate. (It gives us, for example, only a few duds in a retina full of repaired cells.) Suppose further, however, that Cas9 is continuously present in the cells, expressed from the DNA we've delivered. Perhaps over 1 week the errors are 1 in 1000, but over 1000 weeks (20 years) it is close to certain that an off-target edit occurs in each cell. The actual numbers depend on the details of the binding and the mathematics of chance. In some cases, the benefits outweigh the risks; in some, they don't. We could push the odds dramatically in our favor if we could turn Cas9 off when its intended task is complete rather than letting it roam, active, forever.

In the incessant competition between organisms, it seems that every ploy has a counterploy, every weapon is met by a defense, and every defense has a weapon that evolves to subvert it. It perhaps shouldn't surprise us, then, that if CRISPR exists so does anti-CRISPR. Viruses have developed tools to disable the machinery of the bacterial immune system. Anti-CRISPRs were discovered around 2012 as graduate student Joe Bondy-Denomy, in the lab of Alan Davidson at the University of Toronto, noticed that some viruses managed to mount a successful invasion of bacteria that possessed CRISPR sequences against them. Earlier invaders, it turned out, had inserted into the bacterial genome genes for proteins that thwart Cas. There are a multitude of different anti-CRISPRs—over 50 have been discovered so far—targeting every imaginable component of the process sketched above. Some anti-CRISPRs prevent Cas proteins from loading guide RNA. Some cut the guide RNA. Some block the binding of DNA by Cas proteins. Some work by mechanisms yet to be deciphered. In every case, though, biophysical interactions are central. In blocking DNA binding, for example, some anti-CRISPR proteins mimic the electrical charge profile of DNA, mirroring the strength and spatial arrangement of its negatively charged surface to snuggle into the Cas binding groove, rendering it unavailable for its intended DNA. Reminiscent, perhaps, of our applying our knowledge of DNA's electrical properties to move it through gels (chapter 13), viruses make use of these properties for their own ends.

Almost immediately after their discovery, researchers realized that anti-CRISPRs could be crucial additions to the gene editing tool kit. In 2017, Jennifer Doudna and others, including Bondy-Denomy (by this time head of his own lab at the University of California at San Francisco), showed in human cells that delivering an anti-CRISPR after Cas9 effectively shuts down further gene editing and hence reduces unwanted genetic alterations.

In the past few chapters, we've looked at DNA from a broader physical perspective than we applied in part I. Understanding the nature of this molecule and how it interfaces with other materials, whether proteins like polymerases and Cas9 or inorganic structures like the semiconductors of chapter 13, has given us the ability to read and write in the language of genes. In the next and final chapter, I say more about what we can do with this ability, focusing especially on the editing of human genomes and the reshaping of ecosystems, both of which bring issues of ethics to the fore. I also comment on the type of understanding that a biophysical perspective gives us, and its relevance to questions both practical and profound.

Designing the Future

For as long as we've been human, we've sought principles that make sense of the complexity of life. It was widely believed in medieval Europe and the Islamic world, for example, that every species on land has its counterpart in the sea, that Creation had provided "the horse and the sea-horse, the dog and the dog-fish, the snake and the eel," as T. H. White noted in an introduction to a twelfth-century bestiary. This belief has faded—it provides neither descriptive accuracy nor predictive power—but we can sympathize with its hope that some simple symmetry governs living things. Biology often seems a cacophony, beautiful but dizzying. Evolution provides one unifying scaffold, but it primarily illuminates the processes that mold forms over many life spans rather than explaining the connections between form and function.

Over the past several decades, however, we've grown to understand the principles underlying life—physical underpinnings involving self-assembly, randomness, networks of regulatory interactions, and scaling relationships that guide the activities of all creatures. These frameworks connect the microscopic and macroscopic worlds, linking the structure and dynamics of molecules to the workings of cells, tissues, and whole organisms. We could extend our descriptions to still larger scales: communities of species and whole ecosystems are also subject to similar rules. The mathematics of population growth and the vagaries of foraging, for example, tend toward chaotic dynamics with random-looking features that mirror the randomness of their microscopic counterparts. Competition and cooperation lead to self-organizing networks of interactions that govern population sizes. Areas and abundances, as we briefly noted in chapter 12, may obey general scaling relationships. Our focus has mostly been on the scales from molecules to individuals, however, which provide more than enough mystery to keep us occupied.

Understanding biophysical principles brings with it not only a deeper appreciation of life but the ability to reshape it. We've seen several examples of how insights into physical mechanisms intersect issues of health and disease, from the behaviors of pathogenic microbes to the mechanics of organs to risk prediction based on reading DNA. In the next few sections, we look briefly at contemporary examples of modifications to ourselves and our environment that even more dramatically bring to the fore ethical and social concerns. While not central to our aim of describing the workings of life, we'd be remiss to ignore these issues. Moreover, I claim, the biophysical perspective we've developed can help us grapple with difficult questions at the intersection of science and ethics, clarifying the possibilities and impossibilities of technologies that are often popularly described in vague or sensational terms.

Our set of biotechnological tools continues to expand, along with its applications. It is tempting to conclude with a survey of possible future feats—resurrected mastodons and mammoths, an end to animal-derived meat, cures for every disease, engineered plagues, dystopias or utopias of designer babies. The list is limitless, however, and the future is notoriously hard to predict. Instead, we touch on broader themes that, I hope, will be useful regardless of where the future takes us. We end by asking more deeply what our biophysical understanding reveals about the wonders of life, addressing along the way the question of what "understanding" means.

WHAT DOES IT MEAN TO EDIT AN EMBRYO?

The most dramatic new tools are those that edit genomes. We examined in the last chapter therapies related to CRISPR/Cas9 that promise relief from debilitating diseases beyond the reach of conventional treatments. We also saw in chapter 14 that reading the DNA of embryos enables prediction of future traits. It takes little imagination to combine the two and edit embryonic genomes. Delivering CRISPR/ Cas9 to the single cell of a fertilized egg transforms the targeted gene, with the change faithfully copied upon each round of cell division, reaching not only each cell of the fully grown organism but also each of its children, each of their children, and so on. Editing embryos was successfully demonstrated in mice and zebrafish in 2013 and in monkeys in 2014. The latter especially was heralded as a means of generating mimics of human disorders in a similar animal, thereby aiding the development of therapies. It should surprise no one after the preceding chapters that this technique will also work in human embryos—the basic machineries of life are nearly identical in humans as in other animals, to a degree that would have been incomprehensible to our medieval ancestors.

Whether human embryo editing *should* be done, however, is a very different issue. The current consensus is no, at least for embryos that would be implanted to seed viable pregnancies. In fact, such procedures are illegal in dozens of countries, including the United States. The caution arises from the same sorts of ethical concerns highlighted in chapter 14, amplified by the ability of CRISPR/Cas9 to make truly novel changes to genomes rather than just rearranging existing variations. Moreover, solutions to complications such as off-target edits are, as we saw in the last chapter, currently under development. While very promising, few would describe the story as complete. An overwhelming number of scientists, ethicists, and policy makers therefore believe that the time for human embryonic genome editing has not come. Not everyone agrees, however.

In November 2018, Chinese researcher He Jiankui stunned the world with the announcement that his team had used CRISPR/Cas9 to produce the first ever gene-edited babies, a pair of twin girls. Condemnations from around the globe were strong and swift. Chinese authorities quickly shut down He's lab and declared his work "extremely abominable in nature." One can imagine making a case for rushing ahead with human embryo editing if there's imminent danger, perhaps the certainty of a lethal genetic defect, but this was not the case here. The alteration was the disruption of a gene we've seen already, CCR5, whose absence lowers the likelihood of contracting HIV. The choice is puzzling. The girls' father is HIV-positive, but transmission from a father to a child is exceptionally unlikely. The stated aim was to lower the odds of infection during the children's lives, but there are plenty of standard means of effectively preventing HIV. What's more, CCR5 isn't a useless gene. There's good evidence that it helps people resist the impact of other viruses, such as influenza and West Nile virus. Its deletion, therefore, comes with a cost. Even if our target were better justified and the potential benefits outweighed the risks, we should still only proceed if the people involved, namely the parents, are educated about the meaning and consequences of the action. The latter underlies the important ethical principle of informed consent, and it was violated in the He debacle. He's information for the parents was minimal and misleading, and documentation for ethical review procedures was found to be forged. In 2019, He was sentenced to three years in prison.

One could imagine an alternative scenario, in which the first direct alteration of a human embryonic genome occurred after transparent deliberation, and in which the target was an obviously harmful genetic mutation, like the ones that cause muscular dystrophy, cystic fibrosis, or a host of other severe diseases. Such was not the case, however, and despite its technical success, the CCR5 episode will likely muddy the waters of CRISPR-based biotechnologies. The intersection between science and human drama is often messy.

More relevant to our aims here, these events highlight the importance of education. What does "informed consent" mean in the era of genome sequencing and gene editing? At least some awareness of the nature of genes, regulation, probability, and variation, and how all these together orchestrate health and disease, are essential for being an informed participant in any cutting-edge biotechnological therapy. In the absence of understanding, it is all too easy to imagine hopes raised to unrealistic heights, automatic rejection of any genetic technologies, or simple acquiescence to the desires of doctors or the state. Fostering education will be at least as big a challenge as the technical tasks of building organs on a chip or rewriting immune cell genomes. There are precedents, however. The germ theory of disease, for example, has permeated our awareness, with nearly all of humanity possessing a basic mechanistic understanding that microbes exist, can replicate, can infect other creatures, and can cause disease. As noted in chapter 14, almost no one looks in horror at in vitro fertilization; again, a mechanistic understanding of conception has become commonplace. The same awareness can, and should, occur for more recent insights into the workings of life.

THE UNNATURAL WORLD

Modifying plants is contentious. It is unnatural. We've looked at rice and tomatoes; let's now consider carrots, specifically the "baby carrots" you may have seen at a grocery store. They're small, smooth, perhaps even cute, and they account for about 70% of carrot sales in the United States. One might think that they're young carrots, as the name implies; they are not. One might think they are a miniature variety, as ponies are to horses; they are not. Rather, they are full-size, craggy, ugly carrots whittled by machines into small cones. The shaved-off bulk is waste to be discarded. To again use a loaded term, baby carrots are unnatural. If one were to bioengineer an actual miniature carrot, controlling its shape and form, designing cuteness into its genome, it would be unnatural as well. One can wonder which form of unnaturalness would be more palatable, and why.

The existence of baby carrots leads to a more general statement: our world is much less pristine than many people believe. Consider nitrogen, a necessary component of many of the molecules central to life, including DNA and proteins. Nitrogen is an abundant element; as a gas, it makes up about 78% of the atmosphere. This form is inaccessible to animals and plants; none can extract nitrogen from the air. Some bacteria can, however—especially in symbioses with certain plants, they create the nitrogen-containing chemicals that serve as precursors for the molecules with which the plants build themselves and that fertilize the soil after their death, providing nitrogen for other plants and for the animals that eat them. The process works slowly, and insufficient nitrogen is often the limiting factor for the productivity of agricultural land. In the early twentieth century, German chemists Fritz Haber and Carl Bosch developed an artificial process for extracting nitrogen from the air, allowing the industrial production of nitrogen-containing fertilizers. The impact on food production, and civilization, has been enormous. The Haber-Bosch process is the largest single factor underlying humanity's population explosion, and it is estimated that without it, half of the people on the planet would not exist. (The world's population is currently just under 8 billion; in 1900, it was 1.6 billion.) A staggering 50% of the nitrogen atoms in your body, incorporated into DNA strands, amino acid chains, and other materials, are derived from the Haber-Bosch process rather than the natural route of bacterial activity.

The imprint of unnatural processes extends beyond the atoms in our bodies. About 40% of the earth's land is used for agriculture; the percentage is even higher if we exclude barren expanses, like the Sahara and the Antarctic, from the denominator. The atmosphere as well is being reshaped: For the past 10,000 years, its concentration of carbon dioxide ranged between 250 and 280 parts per million. Now, it is over 400, the increase being a consequence of fossil fuel combustion. Through the well-understood physics of thermal radiation, rising carbon dioxide levels lead to an overall increase in the temperature of the planet.

We live, therefore, on a planet in which human activity isn't a small perturbation but is instead a major factor in global systems. Many of us believe that nature is worth preserving, that even aside from its utility there is an intrinsic value to having an awe-inspiring variety of species and landscapes. Preserving nature is, I would argue, paradoxically ill-served by dialing back our unnatural activities. To put it bluntly, that ship has sailed. Eliminating chemical fertilizers and pesticides, for example, with no new technologies brought in to replace them would lead to a roughly fourfold increase in the amount of land required for agriculture. That extra land doesn't exist. The human population continues to grow, and pressures on wilderness and wildlife grow alongside. Bioengineering crops to achieve greater efficiency, however, could enable a reduction in our agricultural footprint, which is hard to imagine occurring by any other means.

ECOSYSTEM ENGINEERING

If accelerating the human manipulation of the living world seems risky, that's because it is. History provides clear instances of ecosystem tinkering gone awry. The story of cane toads provides an example. A century ago, sugar cane crops in Australia were being attacked by beetles. In other parts of the world, cane toads consume beetles and other pests. Why not bring these voracious amphibians, native to Central and South America, down under? And so in the 1930s, cane toads were deliberately introduced to Australia to protect the sugar cane crop. The plan failed catastrophically. The toads gleefully multiplied, with a population now numbering around 200 million and covering a range of about 200,000 square miles (500,000 square kilometers). They ignored the cane beetles, instead gorging on native insects and frogs, bird eggs, and more. Being poisonous, the toads killed many potential predators that might have kept them in check, as well as domestic pets. Rather than being a benefit, the cane toads became a major pest, sending ripples through the continent's habitats. It is still unclear how to deal with them.

Not all plans succeed, but one can hope that we may learn from the past, design better tests and experiments, and consider different tools. In the previous chapter, we looked at CRISPR/Cas9 applied to cells and individual organisms. Here, as our final biotechnological illustration, we see how CRISPR/Cas9 can modify, and even eliminate, an entire species via a construction known as a *gene drive*. Whether or not to use this is a difficult question, as we'll see shortly.

Imagine a trait conferred by a single gene—the common variant, for example, leads to gray insects and a particular mutation in this gene gives black insects. (Or, foreshadowing an application described below, the mutation leads to sterility in mosquitoes if present in both copies of the genome.) Suppose just one individual has the gray mutation. When it mates with a nonmutant, the randomness of recombination gives a 50% chance of the mutation being passed on in the sperm or eggs of the following generation, and unless there's some advantage to being black, the mutation is unlikely to spread through the population.

(I've illustrated this for pairs that always have two offspring, in a schematic that ignores the randomness of inheritance.)

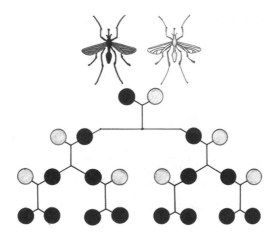

Now suppose instead that, along with the black mutation, this individual's genome has been engineered to contain CRISPR/Cas9, with its target, encoded by the CRISPR sequence, being the "normal" (gray) form of the gene. If the modified DNA encounters an unmodified genome—provided by the unedited mate—Cas9 breaks the gray pigment gene. As we've seen, cells repair broken DNA, and they often use the counterpart provided by the other half of a chromosome pair as a template. The template, containing the black mutation and the CRISPR/Cas9 sequence, is now present in *both* strands. All the future offspring will be black, and will carry the CRISPR/Cas9 machinery that ensures continued edits upon mating with unmodified individuals. The black insects, therefore, will spread through the population.

Why would anyone want to do this? One compelling reason is to eliminate malaria. Each year over 400,000 people die from malaria and more than 200 million are infected, suffering fever, fatigue, and headaches even if they eventually recover. The disease is caused by a parasite transmitted by the female *Anopheles* mosquito, which delivers it to and from human hosts as it drinks their blood. Mosquitoes spread other serious diseases as well, such as dengue fever, West Nile fever, and Zika fever. A variety of government agencies, nonprofit organizations, and public health groups are intensely discussing gene drive targeting of malaria-transmitting *Anopheles* mosquitoes, either by engineering parasite resistance into the insects or by eliminating these mosquitoes entirely. The latter can be done by spreading a mutation that induces maleness, for example, leading to an overwhelmingly male population that can't reproduce, or by disrupting genes related to egg production and therefore spreading sterility.

Though laboratory studies of gene drives in mosquitoes have unfurled as expected, the modified insects have not been released in the wild. However, other genetically altered mosquitoes, modified by older methods that don't propagate through a population, have been set loose. Over a billion male mosquitoes with a genetic variation lethal to their offspring have been set free in Brazil, Malaysia, and the Cayman Islands since 2009. Oxitec, the biotechnology company that developed the insects, notes that their deployment has been successful, causing, for example, an 80% decline in the mosquito population at its Grand Cayman test site and a 90% decline in Jacobina, Brazil. In 2021, the insects were released in the Florida Keys, a region that has seen a surge in cases of dengue fever and Zika. The approval of the Florida plan, the result of a 4–1 vote by the local Mosquito Control District Board of Commissioners in August 2020, followed contentious hearings on whether to adopt the new approach; the standard control method is aerial spraying of pesticides, thought to kill 30%–50% of the mosquitoes.

Intriguingly, Oxitec's modified mosquitoes highlight a biophysical mechanism we've encountered: the lethal variation is not in a gene that encodes some biochemically specific activity, but in a transcription factor (chapter 4), altering the creatures' regulatory circuitry. It seems that the mutant mosquitoes produce an activator that overstimulates its own production, giving a feedback loop of ever-increasing but pointless protein production that jams the insects' cellular machinery. In the lab, males, females, and their progeny can be kept alive by a drug that inhibits the regulatory circuit; this is absent in the wild. Young, rapidly growing mosquitoes are highly sensitive to the regulatory imbalance and die. There are similarities and differences between this approach and that of a gene drive. For example, both involve changes in the genome: the former requires continued introduction of modified individuals, and the latter propagates changes through generations. Understanding these contrasts is important to making informed decisions about how to tackle pressing public health problems like insect-borne diseases.

Returning to gene drives: Their potential deployment in the wild brings trepidation. The elimination of a species could harm its predators or the broader food webs of which it's part, in addition to diminishing the diversity of animal life on the planet. For malaria-linked mosquitoes, the case is stronger than for most applications: the insects aren't a primary food source for other creatures, and even if we erase a few mosquito species, over 3000 other mosquito species remain. And, of course, on the other side of the ethical coin is a vast amount of human suffering. There are still more issues to consider, however. For example, who decides to deploy a gene drive? For mosquitoes and malaria, there is widespread agreement that actions in Africa, home to over 90% of cases, should be decided in Africa. But who in Africa? Mosquitoes don't recognize national borders, and a gene drive propagates to wherever its targets disperse.

These decisions intersect issues of conservation as well. Many fragile ecosystems are threatened by invasive species. For example, rats on the Galápagos Islands plague local wildlife, devouring the eggs of birds and turtles. Would a gene drive for rats be a better method of protecting native species than the rat-killing poisons currently used? Similar questions apply to other regions and other species, including agricultural crops and their pests.

Using gene drives and analogous technologies brings to mind all our past failures of ecosystem engineering, as it should. One worries that we will simply see new versions of the cane toad saga. We can learn from our mistakes, though, and also take heart that our current technologies are more precise and focused than those of the past. Additional enhancements that offer safeguards are already under development, such as DNA-encoded mechanisms to block or even reverse gene drives in a population. Surgery used to be crude and often lethal due to exposure to infection; now it is routinely successful thanks to better insights and methods. It is not unreasonable, I claim, to work toward similarly effective ends for larger-scale interventions. This may seem unduly optimistic, but, as noted above, the alternatives to embracing modern methods may be even less palatable.

There's much more we could write about the intersection of biotechnology and society. New developments and implementations, exhilarating as well as risky, will continue to arrive not only because of the depth of our insights into the living world but because the tools we're creating are cheap and easy to use. I see these days advertisements offering to sequence my entire genome for $300, a far cry from the $3 billion of 30 years ago or even the $10,000 of 10 years ago. Machines for performing PCR, purifying proteins, and growing cells are more accessible than ever. In fact, communities of amateur enthusiasts share biochemical recipes and designs for 3-D-printed equipment with which to explore biotechnological crafts from their garages. Members of the Open Insulin Project, for example, aim to develop "freely available, open organisms for insulin production" that will liberate diabetics from commercial insulin suppliers.

One might be aghast that biotechnological paraphernalia is so readily available, or delighted at its democratization, but regardless of our response, the ubiquity of such tools is a reality. The technologies that form the basis of futuristic applications exist now, and they make possible such widespread and uncontroversial applications, like diagnosing disease and monitoring microbial contaminants, that it is hard to imagine we will stop using them or stop thinking of new targets at which to direct them. I've often used the word *we* when writing, glossing over who exactly "we" refers to. In the context of biotechnology, for better or worse, "we" can be anyone. Any reasonably advanced country, or even institution, is capable of using the tools we've encountered in this book. We'll see how these instruments are put to use in the years to come. Once again, I claim, education is crucial to ensuring that the decisions we make are the best they can be, informed by an understanding of how the technologies at our disposal work and what they can and cannot do. Education, of course, has benefits beyond the practical. We end by stepping back to the deeper, and more inspirational, lessons drawn from looking into the workings of life.

UNDERSTANDING UNDERSTANDING

The most important theme of this book is that themes exist. The living world isn't just a collection of anatomical structures and biochemical reactions. There are principles, motifs, and mechanisms that unify all of its components and their activities. This statement shouldn't come as a surprise after the preceding 15 chapters. I've avoided until now, however, the question of whether this unity matters, either practically or aesthetically. This question underlies a tension present in much of contemporary science that I illustrate with a nonbiological example before we return to the living world.

There wasn't a singular experience that drove me to study physics, but one event I remember vividly occurred in my high school physics class, an activity in which we released a ball from a ramp onto a table, along which it rolled until plummeting off the table's edge to fall into a plastic cup on the floor. Our task: to predict beforehand where to place the cup to catch the ball. It may seem dull on paper, but to see that the ball did, in fact, land exactly where universal laws of motion say it should was amazing to me. We needn't memorize facts about balls, ramps, or cups, or even care about them; instead, there are general themes that guide us. Searching for and applying broad principles is central to physics and underpins the biophysical approach that I've championed throughout this book.

There is, however, another way we could have made our prediction. From prior experiments, we could have tabulated a vast list of outcomes. For a range of ramp heights, tilt angles, ball masses, ball materials, table heights, air temperatures, and whatever other parameters might possibly be relevant, we could have released balls and recorded where they landed. Then, for the conditions of our new roll, we could look up the outcome from our giant database. This, to use the current jargon, is the "big data" approach. There's nothing wrong with it, and it's often very successful. We've seen this already, in the correlations observed between genomic features and traits such as human height (chapter 14). These correlations enable predictions, but we have no understanding of why the correlations exist.

The contrast between understanding basic principles and cataloging information arises with increasing frequency throughout science, as our tools for generating and processing data become more powerful. Sometimes, as in the applications of Brownian motion (chapter 6), the basic principles are so profound and well understood that it would be absurd to ignore them. Sometimes, as in the genomics example above, looking for correlations is the only possible approach, at least for now. Sometimes it is unclear what to do. In chapter 4, we looked at the biological circuits that regulate genes, composed of repressor and promoter proteins that orchestrate gene activity. To some researchers, identifying basic circuit motifs—feedback loops, oscillators, clocks, and so on—is the path to progress. To others, this is irrelevant; plugging in the details of all the interactions between all the components of a regulatory network, even if they number in the hundreds or thousands, is the more useful strategy. It isn't obvious whether the former, more biophysical approach or the latter, more phenomenological approach is better. They aren't mutually exclusive; pursuing both in parallel may be the best plan, especially as the detailed models might help uncover core concepts that are yet unknown. Still, the tension between these philosophies is real, and it colors debates about funding and research directions.

My own bias is toward illuminating basic, minimal sets of principles. After all, a giant database of falling ball trajectories would enable prediction of future experiments on falling balls, but it's the deeper understanding of Newton's laws of motion that gave us the trajectories that sent astronauts to the moon. Which approach is the more practical, therefore, may depend on the timescale: immediate application to some problem at hand, or to problems of the future that are still unimagined. Aside from the aims of practicality, there's also the deeply human appeal of understanding. *Understanding* is hard to define. Still, extracting from complexity simple yet powerful explanations has compelled us for as long we've been human, powering myths as well as science.

TWO ENDINGS TO OUR STORY

There is a way one may have expected this book would end. We began with the ingredients of life, molecules and machineries that generate the inner dynamics of cells. We then examined communities of cells, as in organs and embryos, and then some of the principles that guide larger-scale shape and form. Finally, we reconnected with the subcellular world, learning how to read and write the molecular code of the genome. One might think, therefore, that the loop is closed and that we could conclude by describing the procedure for designing organisms at will—perhaps nutritious, drought-resistant plants, or animals resurrected from human-induced annihilation—by crafting and executing the code that gives creatures of whatever body and behavior we want. For example, we would sketch how to transform a tiny dik-dik antelope into a massive wildebeest, poring over the clues in its DNA, nudging the nucleotide letters in the appropriate genes to spin new amino acid chains that fold into new loops and sheets, guiding the expression of other genes and forming the substance of bones and muscle, eyes and lungs, all designed to accommodate the pull of the earth, the inhalation of air, and the demands of the environment.

We do not, however, know how to design such a transformation, and this is not how the book ends. Our ignorance isn't a sign of failure;

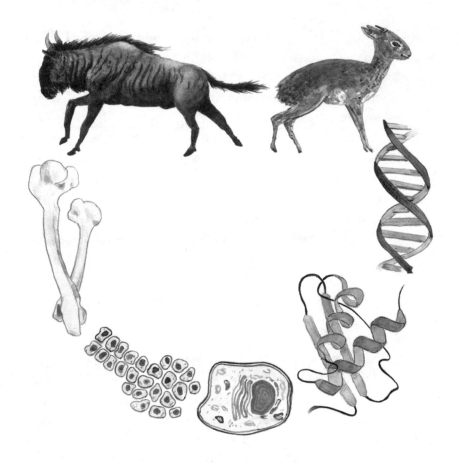

rather, it tells us that the story we've been exploring is far from finished. We know the basic pieces and principles that make up life. Our
grasp of the detailed contexts in which they are manifested, however,
is small, but growing. It is as if, having recently learned to read, a whole
library awaits us. It is not certain that we will be able to make sense of
it all. It may be that, though the general themes are clear, the minu-
tiae of genetic interactions, chemical reactions, and microscopic forces
that give rise to bamboo rather than beech, that separate the horse and
the hippo, and that set the precise ratio of femur width to length will
be too difficult to entirely comprehend or may not be worth the effort.
This remains to be seen. Even if so, it would not diminish the accom-
plishment of understanding the architecture of life, nor dampen our
enthusiasm for further exploration.

The fundamentally human urge to know, to seek connections among phenomena and to feel joy when we find such connections, provides the most powerful motivation for studying nature. Especially in the past few chapters, we've looked a lot at technological applications of our modern understanding of life. Technology and its intersection with health, disease, and society are important, and to many people they're the most important facet of science. I would study biophysics, however, and I would have written this book even if these applications didn't exist. The living world is full of an awe-inspiring variety of forms and activities. In our own backyards and urban parks, squirrels leap among branches, sunlight glints off the diaphanous wings of dragonflies, and trees pull carbon from the air to build towers tens of feet tall. In more exotic locales, lions stalk the savannas, dolphins play among waves, and fish miles deep in ocean trenches generate their own light. We can see all of this and be struck by its beauty. We can, in addition, now appreciate the miles of DNA packed inside the lion, the dance of neurotransmitters as it decides to pounce, the march of motor proteins along the fibers of its cells—layers of phenomena as remarkable as those that are more obvious. What's more, we can contemplate a unity among all creatures, including ourselves, beyond the dreams of our ancestors. Whether on land or sea, newborn or grown, microscopic or gigantic, every organism is made of the same molecular building blocks, a handful of molecules that encode information and build themselves into three-dimensional shapes. Every organism is shaped with the physical forces of the universe; the bones of both the lion and the antelope she stalks are governed by the constraints of gravity, and the dolphin could only be a dolphin in a large liquid home. Every organism not only tolerates the microscopic chaos of jostling molecules but transforms the chaos into computation, regulating the activity of genes, proteins, cells, and organs in response to stimuli from within and without. Beyond superficial appearances, as wonderful as they are, is a profound and elegant framework that makes life work and that we now begin to appreciate.

Acknowledgments

The idea of writing this book dates back a decade. Around the same time, I created a course called "The Physics of Life," with the aims of conveying the wonders of biophysics to non-science-major undergraduates and using biophysics as a tool to foster scientific literacy. This book is much broader than the course and is structured quite differently, but nonetheless I'd like to thank the course's many students not only for their frequent enthusiasm but also for their moments of polite but stone-faced indifference. As a scientist, one tends to find fascinating almost everything even tangentially connected to one's field; it can be shocking to learn that some topics are, to most people, quite dull. The spur to constantly try different tactics to enliven certain subjects, or to drop them entirely, was invaluable. I'd also like to thank the University of Oregon's Science Literacy Program, especially Elly Vandegrift, Michael Raymer, and Judith Eisen, for creating a program that encourages thinking deeply about science communication.

I'm grateful to Sergio, Miguel, Pablo, Jesse, Chilo, and everyone else who works at Espresso Roma, along with its owners, Miguel and Maria Cortez, for great coffee and a wonderful environment in which to think and write.

David Rabuka and Phil Nelson read drafts of every chapter and generously provided extensive, enthusiastic, and insightful comments. I am delighted to be able to thank them in print. Phil furthermore helped launch this project, not only with words of encouragement but also by serving as matchmaker with my editor-to-be, Jessica Yao of Princeton University Press. To Jessica, warm thanks not only for championing this book but also for an abundance of perceptive suggestions. Sincere thanks also go to Ingrid Gnerlich, the project's enthusiastic second editor.

Finally, I am deeply grateful to my wife, Julie, and to the kids, Kiran and Suryan, who are always wonderful and who put up with the many hours I've spent distracted by reworking sentences and drawing DNA.

References

GENERAL

Standard biology textbooks cover many well-established facts that are especially relevant to part I. See, for example:

- B. Alberts, A. D. Johnson, J. Lewis, D. Morgan, M. Raff, K. Roberts, P. Walter, *Molecular Biology of the Cell* (W. W. Norton, New York, sixth edition, 2014). The fourth edition is freely available: http://www.ncbi.nlm.nih.gov/bookshelf/br.fcgi?book=mboc4.

Excellent textbooks on biophysics, aimed at advanced undergraduates or graduate students, include

- P. Nelson, *Biological Physics: Energy, Information, Life* (Chiliagon Science, Philadelphia, student edition, 2020).
- R. Phillips, J. Kondev, J. Theriot, H. Garcia, *Physical Biology of the Cell* (Garland Science, London, second edition, 2012).
- W. Bialek, *Biophysics: Searching for Principles* (Princeton University Press, Princeton, NJ, 2012).

INTRODUCTION

p. 1 "Ancient Indian texts applied a variety of classifiers"

- B. K. Smith, Classifying animals and humans in ancient India. *Man* **26**, 527–548 (1991).

p. 6 "arrangements of adjoining cells in all sorts of tissues resemble the arrangements of soap bubbles"

- D. W. Thompson, *On Growth and Form* (Cambridge University Press, Cambridge, UK, 1942).

p. 6 "Takashi Hayashi . . . and Richard Carthew . . . looked at the cluster of photoreceptor cells situated in each of a fruit fly's compound eyes"

- T. Hayashi, R. W. Carthew, Surface mechanics mediate pattern formation in the developing retina. *Nature* **431**, 647–652 (2004).

p. 7 "assessing the mechanical stiffness of the neighboring tissue"

- E. Puklin-Faucher, M. P. Sheetz, The mechanical integrin cycle. *Journal of Cell Science* **122**, 179–186 (2009).

- A. del Rio, R. Perez-Jimenez, R. Liu, et al., Stretching single talin rod molecules activates vinculin binding. *Science* **323**, 638–641 (2009).

p. 9 "In the twentieth century alone, for example, more than 300 million people died of smallpox"

- D. A. Henderson, The eradication of smallpox—an overview of the past, present, and future. *Vaccine* **29**, D7–D9 (2011).

p. 11 "At the end of *On the Origin of Species*, Darwin writes"

- C. Darwin, *On the Origin of Species by Means of Natural Selection, or, the Preservation of Favoured Races in the Struggle for Life* (J. Murray, London, 1859), 490.

CHAPTER 1. DNA: A CODE AND A CORD

p. 15 "A beige gelatinous slab speckled with bacterial colonies hangs in the National Portrait Gallery"

- D. Nelkin and M. S. Lindee, *The DNA Mystique: The Gene as a Cultural Icon* (W. H. Freeman, 1995).

p. 15 "it carries the actual instructions that led to the creation of John"

- BBC News, "Gallery puts DNA in the frame," September 19, 2001. http://news.bbc.co.uk/2/hi/entertainment/1550864.stm.

p. 20 "James Watson and Francis Crick figured out the structure of double-stranded DNA." Much has been written about this history, and the roles of Watson, Crick, Rosalind Franklin, and others. See, for example:

- J. D. Watson, *The Double Helix: A Personal Account of the Discovery of the Structure of DNA* (Atheneum Press, New York, 1968).
- A. Sayre, *Rosalind Franklin and DNA* (W. W. Norton, New York, 2000).
- M. Cobb, Sexism in science: Did Watson and Crick really steal Rosalind Franklin's data? *The Guardian*, June 23, 2015. http://www.theguardian.com/science/2015/jun/23/sexism-in-science-did-watson-and-crick-really-steal-rosalind-franklins-data.
- S. Mukherjee, *The Gene: An Intimate History* (Scribner, New York, 2016).

p. 23 On phase transitions in general:

- N. Goldenfeld, *Lectures on Phase Transitions and the Renormalization Group* (Addison-Wesley, Reading, MA, 1972).

pp. 23–24 On the physics of DNA melting:

- M. Peyrard, Biophysics: Melting the double helix. *Nat. Phys.* **2**, 13–14 (2006).
- M. Peyrard, Nonlinear dynamics and statistical physics of DNA. *Nonlinearity* **17**, R1–R40 (2004).

p. 24 "In 1983, the recipe that combines nucleotides, polymerases, temperature, and DNA came to scientist Kary Mullis." A first-person account of the discovery of the polymerase chain reaction:

- K. B. Mullis, The unusual origin of the polymerase chain reaction. *Sci. Am.* **262**, 56–65 (1990).

CHAPTER 2. PROTEINS: MOLECULAR ORIGAMI

p. 30 On the history of X-ray crystallography:

- J. C. Brooks-Bartlett, E. F. Garman, The Nobel science: One hundred years of crystallography. *Interdisciplinary Science Reviews* **40**, 244–264 (2015).
- M. Jaskolski, Z. Dauter, A. Wlodawer, A brief history of macromolecular crystallography, illustrated by a family tree and its Nobel fruits. *FEBS Journal* **281**, 3985–4009 (2014).

pp. 30–31 "Kendrew's team struggled with myoglobin from porpoises, penguins, seals, and other creatures"; and the quote from Kendrew, "the most marvelous . . . gigantic crystals"

- S. de Chadarevian, John Kendrew and myoglobin: Protein structure determination in the 1950s. *Protein Science* **27**, 1136–1143 (2018).

p. 32 On green fluorescent protein and other fluorescent proteins:

- R. N. Day, M. W. Davidson, The fluorescent protein palette: Tools for cellular imaging. *Chem. Soc. Rev.* **38**, 2887–2921 (2009).
- E. A. Rodriguez, R. E. Campbell, J. Y. Lin, et al., The growing and glowing toolbox of fluorescent and photoactive proteins. *Trends in Biochemical Sciences* **42**, 111–129 (2017).
- N. C. Shaner, The mFruit collection of monomeric fluorescent proteins. *Clin. Chem.* **59**, 440–441 (2013).

p. 33 The drawing of green fluorescent protein is based on structure 1EMA in the Protein Data Bank:

- https://www.rcsb.org/structure/1EMA.
- M. Ormö, A. B. Cubitt, K. Kallio, et al., Crystal structure of the *Aequorea victoria* green fluorescent protein. *Science* **273**, 1392–1395 (1996).

p. 33 On the structure of potassium channels:

- Q. Kuang, P. Purhonen, H. Hebert, Structure of potassium channels. *Cell Mol. Life Sci.* **72**, 3677–3693 (2015).

p. 33 The drawing of a potassium channel is based on structure 1BL8 in the Protein Data Bank:

- https://www.rcsb.org/structure/1BL8.

- D. A. Doyle, J. Morais Cabral, R. A. Pfuetzner, et al., The structure of the potassium channel: Molecular basis of K+ conduction and selectivity. *Science* **280**, 69–77 (1998).

p. 33 The drawing of kinesin is based on structure 1N6M in the Protein Data Bank:
- https://www.rcsb.org/structure/1N6M.
- M. Yun, C. E. Bronner, C.-G. Park, et al., Rotation of the stalk/neck and one head in a new crystal structure of the kinesin motor protein, Ncd. *EMBO Journal* **22**, 5382–5389 (2003).

p. 34 The drawing of the glucocorticoid receptor is based on structure 1R4O in the Protein Data Bank:
- https://www.rcsb.org/structure/1R4O.
- B. F. Luisi, W. X. Xu, Z. Otwinowski, et al., Crystallographic analysis of the interaction of the glucocorticoid receptor with DNA. *Nature* **352**, 497–505 (1991).

p. 36 "a group of other proteins called *chaperones* comes to their aid"
- F. U. Hartl, M. Hayer-Hartl, Molecular chaperones in the cytosol: From nascent chain to folded protein. *Science* **295**, 1852–1858 (2002).

pp. 39–40 On kuru:
- M. P. Alpers. The epidemiology of kuru: Monitoring the epidemic from its peak to its end. *Philosophical Transactions of the Royal Society London, B: Biological Sciences* **363**, 3707–3713 (2008).
- J. T. Whitfield, W. H. Pako, J. Collinge, M. P. Alpers, Mortuary rites of the South Fore and kuru. *Philosophical Transactions of the Royal Society London, B: Biological Sciences* **363**, 3721–3724 (2008).
- R. E. Bichell, When people ate people, a strange disease emerged. NPR .org, September 6, 2016. https://www.npr.org/sections/thesalt/2016/09 /06/482952588/when-people-ate-people-a-strange-disease-emerged.

p. 40 On prions and their discovery:
- S. B. Pruisner, The prion diseases. *Scientific American* **272**, 48–57 (1995).
- NobelPrize.org, Nobel Media AB press release (1997). https://www .nobelprize.org/prizes/medicine/1997/press-release/.

p. 40 On meat-and-bone meal:
- C. Ducrot, M. Paul, D. Calavas, BSE risk and the use of meat and bone meal in the feed industry: Perspectives in the context of relaxing control measures. *Natures Sciences Societes* **21**, 3–12 (2013).

pp. 41–42 On predicting protein structures:
- F. M. Richards, The protein folding problem. *Scientific American* **264**, 54–63 (1992).
- K. A. Dill, J. L. MacCallum, The protein-folding problem, 50 years on. *Science* **338**, 1042–1046 (2012).

p. 42 "to commission bespoke supercomputers devoted to the biophysical challenge of protein folding"

- A. Elliot, David E. Shaw's supercomputer is uncovering secrets of human biology. Columbia Engineering, April 7, 2017. https://engineering.columbia .edu/news/engineering-icons-david-shaw.

p. 42 "the 'folding@home' program that runs in the background of volunteers' computers"

- Folding@home—fighting disease with a world wide distributed super computer. https://foldingathome.org/.
- K. Greene, 2002, Folding@home takes to the lab. *Science*, October 21, 2002. https://www.sciencemag.org/news/2002/10/foldinghome-takes -lab.

p. 42 "a free protein folding game"

- S. Cooper, F. Khatib, A. Treuille, et al., Predicting protein structures with a multiplayer online game. *Nature* **466**, 756–760 (2010).

p. 42 "DeepMind, a company affiliated with Google"

- R. F. Service, "The game has changed." AI triumphs at protein folding. *Science* **370**, 1144–1145 (2020).
- E. Callaway, "It will change everything": DeepMind's AI makes gigantic leap in solving protein structures. *Nature* **588**, 203–204 (2020).

CHAPTER 3. GENES AND THE MECHANICS OF DNA

p. 46 "Differences of even a single three-nucleotide group . . . can lead to subtle but measurable shifts in color perception . . . several forms of color blindness"

- S. S. Deeb, The molecular basis of variation in human color vision. *Clin. Genet.* **67**, 369–377 (2005).
- MedlinePlus, Color vision deficiency. https://medlineplus.gov/genetics /condition/color-vision-deficiency/(reviewed January 1, 2015; updated August 18, 2020).

p. 47 "An RNA called 'Growth arrest-specific 5'" and other aspects of non-coding RNA:

- T. Kino, D. E. Hurt, T. Ichijo, et al., Noncoding RNA gas5 is a growth arrest- and starvation-associated repressor of the glucocorticoid receptor. *Sci. Signal* **3**, ra8 (2010).
- M. Guttman, J. L. Rinn, Modular regulatory principles of large noncoding RNAs. *Nature* **482**, 339–346 (2012).

pp. 47–48 "The bacteria that cause tuberculosis and cholera each have about 4000 genes in their genome"

- *Kyoto Encyclopedia of Genes and Genomes* (KEGG). https://www
  .genome.jp/kegg/; for *Mycobacterium tuberculosis*, https://www.genome
  .jp/kegg-bin/show_organism?org=mtu; for *Vibrio cholerae,* http://www
  .genome.jp/kegg-bin/show_organism?org=vch (accessed January 14,
  2021).

p. 48 "The genome of the *Lactobacillus delbrueckii* subspecies commonly employed to turn milk into yogurt has about 2000 protein-coding genes"

- National Center for Biotechnology Information. https://www.ncbi.nlm
  .nih.gov/genome/?term=Lactobacillus%20delbrueckii[Organism] (accessed January 14, 2021).

p. 48 "The human genome contains about 20,000 protein-coding genes." For this, and the number of noncoding genes, see

- I. Ezkurdia, D. Juan, J. M. Rodriguez, et al., Multiple evidence strands suggest that there may be as few as 19 000 human protein-coding genes. *Hum. Mol. Genet.* **23**, 5866–5878 (2014).
- C. Willyard, New human gene tally reignites debate. *Nature* **558**, 354–355 (2018).

pp. 48–49 On genomes of various organisms:

- D. M. Church, L. Goodstadt, L. W. Hillier, et al., Lineage-specific biology revealed by a finished genome assembly of the mouse. *PLOS Biology* **7**, e1000112 (2009).
- C. M. Wade, E. Giulotto, S. Sigurdsson, et al., Genome sequence, comparative analysis, and population genetics of the domestic horse. *Science* **326**, 865–867 (2009).
- X. Zhan, S. Pan, J. Wang, et al., Peregrine and saker falcon genome sequences provide insights into evolution of a predatory lifestyle. *Nature Genetics* **45**, 563–566 (2013).
- R. A. Ohm, J. F. de Jong, L. G. Lugones, et al., Genome sequence of the model mushroom Schizophyllum commune. *Nature Biotechnology* **28**, 957–963 (2010).
- J. K. Colbourne, M. E. Pfrender, D. Gilbert, et al., The ecoresponsive genome of *Daphnia pulex. Science* **331**, 555–561 (2011).
- Rice Annotation Project database (RAP-DB): 2008 update. *Nucleic Acids Res.* **36**, D1028–D1033 (2008).
- Gramene database. http://ensembl.gramene.org/Zea_mays/Info/Annotation/.

p. 49 On noncoding genes in rice and maize:

- H. Wang, Q.-W. Niu, H.-W. Wu, et al., Analysis of non-coding transcriptome in rice and maize uncovers roles of conserved lncRNAs associated with agriculture traits. *Plant J.* **84**, 404–416 (2015).

- L. Li, S. R. Eichten, R. Shimizu, et al., Genome-wide discovery and characterization of maize long non-coding RNAs. *Genome Biology* **15**, R40 (2014).

pp. 49–50 On genome sizes:

- The BioNumbers database, http://bionumbers.hms.harvard.edu/search .aspx, contains information on genome sizes, including those of the bread mold *Neurospora crassa* and the soil-dwelling amoeba *Dictyostelium discoideum*; search "number of genes" or the species name.

pp. 54–56 On DNA packaging, histones, and chromatin:

- H. Schiessel, The physics of chromatin. *J. Phys. Condens. Matter* **15**, R699–R774 (2003).
- D. J. Tremethick, Higher-order structures of chromatin: The elusive 30 nm fiber. *Cell* **128**, 651–654 (2007).

p. 54 The drawing of DNA wrapped around a histone complex is based on structure 1AOI in the Protein Data Bank:

- https://www.rcsb.org/structure/1AOI.
- K. Luger, A. W. Mader, R. K. Richmond, et al., *Nature* **389**, 251–260 (1997).

p. 55 "researchers at the Salk Institute . . . developed a method to stain DNA in intact nuclei, decorating them with metal atoms that are readily visible in an electron microscope"

- H. D. Ou, S. Phan, T. J. Deerinck, C. C. O'Shea, ChromEMT: Visualizing 3D chromatin structure and compaction in interphase and mitotic cells. *Science* **357**, eaag0025 (2017).

p. 56 On DNA packaging and disease:

- C. Mirabella, B. M. Foster, T. Bartke, Chromatin deregulation in disease. *Chromosoma* **125**, 75–93 (2016).
- A. DeLaurier, Y. Nakamura, I. Braasch, Histone deacetylase-4 is required during early cranial neural crest development for generation of the zebrafish palatal skeleton. *BMC Developmental Biology* **12**, 16 (2012).

p. 56 "the determinants of what DNA regions are wrapped around histones"

- E. Segal, Y. Fondufe-Mittendorf, L. Chen, et al., A genomic code for nucleosome positioning. *Nature* **442**, 772–778 (2006).
- F. G. Brunet, B. Audit, G. Drillon, et al., Evidence for DNA sequence encoding of an accessible nucleosomal array across vertebrates. *Biophysical Journal* **114**, 2308–2316 (2018).

pp. 57–59 On the packaging of DNA in viruses:

- A. Evilevitch, L. Lavelle, C. M. Knobler, et al., Osmotic pressure inhibition of DNA ejection from phage. *Proc. Natl. Acad. Sci.* **100**, 9292–9295 (2003).

- W. M. Gelbart and C. M. Knobler, Virology: Pressurized viruses. *Science* **323**, 1682–1683 (2009).

CHAPTER 4. THE CHOREOGRAPHY OF GENES

p. 63 The illustration of the lac repressor bound to DNA is based on structures 1EFA and 1TLF in the Protein Data Bank and a composite illustration by David Goodsell:
- https://www.rcsb.org/structure/1EFA.
- https://www.rcsb.org/structure/TLF.
- D. Goodsell, http://pdb101.rcsb.org/motm/39.
- A. M. Friedman, T. O. Fischmann, T. A. Steitz, Crystal structure of lac repressor core tetramer and its implications for DNA looping, *Science* **268**, 1721–1727 (1995).
- C. E. Bell, M. Lewis, A closer view of the conformation of the lac repressor bound to operator. *Nat. Struct. Biol.* **7**, 209–214 (2000).

p. 63 "The protein must therefore loop the DNA into a tight circle"
- R. Schleif, DNA Looping. *Annual Review of Biochemistry.* **61**, 199–223 (1992).

p. 63 "The looped DNA interferes with the normal binding of RNA polymerase"
- Z. Vörös, Y. Yan, D. T. Kovari, et al., Proteins mediating DNA loops effectively block transcription. *Protein Sci.* **26**, 1427–1438 (2017).
- N. A. Becker, J. P. Peters, L. J. Maher, T. A. Lionberger, Mechanism of promoter repression by lac repressor-DNA loops. *Nucleic Acids Res.* **41**, 156–166 (2013).

p. 64 "This phenomenon was discovered in the 1940s by Jacques Monod." For more of the history, see
- M. Morange, *A History of Molecular Biology* (Harvard University Press, Cambridge, MA, 2000).

p. 65 "the human genome contains over 1500 genes for transcription factors"
- S. A. Lambert, A. Jolma, L. F. Campitelli, et al., The human transcription factors. *Cell* **172**, 650–665 (2018).

p. 65 "a transcription factor bound to a segment of DNA can influence expression of a gene that would be distant if the DNA were laid out in a straight line"
- S. Schoenfelder, P. Fraser, Long-range enhancer–promoter contacts in gene expression control. *Nature Reviews Genetics* **20**, 437–455 (2019).

p. 66 "a paper from 2001 by Heidi Scrable and colleagues at the University of Virginia"
- C. A. Cronin, W. Gluba, H. Scrable, The lac operator-repressor system is functional in the mouse. *Genes Dev.* **15**, 1506–1517 (2001).

p. 68 "As Monod himself dramatically and presciently noted"
- H. C. Friedmann, From "butyribacterium" to "E. coli": An essay on unity in biochemistry. *Perspect. Biol. Med.* **47**, 47–66 (2004).

pp. 68–73 On memories, clocks, and other genetic circuits:
- P. C. Nelson, *Physical Models of Living Systems* (W. H. Freeman, 2015).
- U. Alon, *An Introduction to Systems Biology: Design Principles of Biological Circuits* (CRC Press, Boca Raton, FL, 2007).

pp. 70–72 On the circadian rhythm and its cellular clock:
- S. A. Brown, E. Kowalska, R. Dallmann, (Re)inventing the circadian feedback loop. *Dev. Cell* **22**, 477–487 (2012).
- E. S. Maywood, L. Drynan, J. E. Chesham, et al., Analysis of core circadian feedback loop in suprachiasmatic nucleus of mCry1-luc transgenic reporter mouse. *Proc. Natl. Acad. Sci.* **110**, 9547–9552 (2013).
- J. P. Pett, A. Korenčič, F. Wesener, et al., Feedback loops of the mammalian circadian clock constitute repressilator. *PLOS Comput. Biol.* **12**, e1005266 (2016).
- D. R. Weaver, The suprachiasmatic nucleus: A 25-year retrospective. *Journal of Biological Rhythms* (2016).

p. 72 On engineering the repressilator into cells:
- M. B. Elowitz, S. Leibler, A synthetic oscillatory network of transcriptional regulators. *Nature* **403**, 335–338 (2000).

p. 72 "a variety of other precise and tunable oscillators have been engineered into cells." See, for example:
- J. Stricker, S. Cookson, M. R. Bennett, et al., A fast, robust and tunable synthetic gene oscillator. *Nature* **456**, 516–519 (2008).

p. 73 On histone modification:
- M. Lawrence, S. Daujat, R. Schneider, Lateral thinking: How histone modifications regulate gene expression. *Trends in Genetics* **32**, 42–56 (2016).

p. 73 "This modification of histones is particularly important during the early development of an embryo"
- L. Ho, G. R. Crabtree, Chromatin remodelling during development. *Nature* **463**, 474–484 (2010).

p. 74 On epigenetics:
- C. D. Allis, T. Jenuwein, The molecular hallmarks of epigenetic control. *Nature Reviews Genetics* **17**, 487–500 (2016).
- Bošković, O. J. Rando, Transgenerational epigenetic inheritance. *Annual Review of Genetics* **52**, 21–41 (2018).
- B. T. Heijmans, E. W. Tobi, A. D. Stein, et al., Persistent epigenetic differences associated with prenatal exposure to famine in humans. *PNAS* **105**, 17046–17049 (2008).

p. 74 On studies of the 1944–45 Dutch famine survivors:

- R. C. Painter, C. Osmond, P. Gluckman, et al., Transgenerational effects of prenatal exposure to the Dutch famine on neonatal adiposity and health in later life. *BJOG: An International Journal of Obstetrics & Gynaecology* **115**, 1243–1249 (2008).
- M.V.E. Veenendaal, R. C. Painter, S. de Rooij, et al., Transgenerational effects of prenatal exposure to the 1944–45 Dutch famine. *BJOG: An International Journal of Obstetrics & Gynaecology* **120**, 548–554 (2013).

CHAPTER 5. MEMBRANES: A LIQUID SKIN

p. 76 "These membrane-associated proteins account for over a third of the human genome." An even larger number are peripherally associated with the membrane:

- L. Dobson, I. Reményi, G. E. Tusnády, The human transmembrane proteome. *Biol. Direct* **10**, 31 (2015).
- M. S. Almén, K. J. Nordström, R. Fredriksson, H. B. Schiöth, Mapping the human membrane proteome: A majority of the human membrane proteins can be classified according to function and evolutionary origin. *BMC Biology* **7**, 50 (2009).

p. 78 "interactions between two cell types known as T-cells and antigen presenting cells"

- A. Grakoui, S. K. Bromley, C. Sumen, et al., The immunological synapse: A molecular machine controlling T cell activation. *Science* **285**, 221–227 (1999).
- S. K. Bromley, W. R. Burack, K. G. Johnson, et al., The immunological synapse. *Annu. Rev. Immunol.* **19**, 375–396 (2001).

p. 79 "similar synapses form at the contacts between immune cells transmitting the human T-cell leukemia virus as well as the human immunodeficiency virus"

- V. Piguet, Q. Sattentau, Dangerous liaisons at the virological synapse. *J. Clin. Invest.* **114**, 605–610 (2004).

p. 81 "In London at the start of the nineteenth century, 30% of all deaths were due to tuberculosis"

- S. A. Waksman, *The Conquest of Tuberculosis* (University of California Press, Berkeley, 1964).

p. 81 "remained the first or second leading cause of death each year in the United States through the first decade and a half of the twentieth century"

- Centers for Disease Control (USA), Leading causes of death, 1900–1998. https://www.cdc.gov/nchs/data/dvs/lead1900_98.pdf.

p. 81 "Even now, about one million people die of tuberculosis annually"
- World Health Organization, WHO global tuberculosis report, 2017. http://www.who.int/tb/publications/global_report/en/.

p. 81 "We've known for about a century, for example, that *Mycobacterium leprae* and *Mycobacterium tuberculosis* can survive periods of dehydration lasting several months"
- D. C. Twitchell, The vitaility of tubercle bacilli in sputum. *Transactions of the National Association for the Study and Prevention of Tuberculosis, Annual Meeting*, 221–230 (1905).
- M. B. Soparker, The vitailty of tubercle bacilli outside the body. *Indian J. Med. Res.* **4**, 627–650 (1917).
- C. R. Smith, Survival of tubercle bacilli. *American Review of Tuberculosis* **45**, 334–345 (1942).

p. 82 On the use of trehalose for dehydration resistance:
- J. H. Crowe, F. A. Hoekstra, L. M. Crowe, Anhydrobiosis. *Annu. Rev. Physiol.* **54**, 579–599 (1992).

pp. 82–84 My lab's work on trehalose-containing lipids:
- C. W. Harland, D. Rabuka, C. R. Bertozzi, R. Parthasarathy, The *M. tuberculosis* virulence factor trehalose dimycolate imparts desiccation resistance to model mycobacterial membranes. *Biophys. J.* **94**, 4718–4724 (2008).
- C. W. Harland, Z. Botyanszki, D. Rabuka, et al., Synthetic trehalose glycolipids confer desiccation resistance to supported lipid monolayers. *Langmuir* **25**, 5193–5198 (2009).

p. 87 "chemically perturb cells to create 'blebs' . . . . In these, one finds visibly discernible lipid phases"
- T. Baumgart, A. T. Hammond, P. Sengupta, et al., Large-scale fluid/fluid phase separation of proteins and lipids in giant plasma membrane vesicles. *Proc. Natl. Acad. Sci.* **104**, 3165–3170 (2007).

p. 87 "scientists have observed large, visible domains in the membranes that bound an organelle called the *vacuole* in yeast cells"
- S. P. Rayermann, G. E. Rayermann, C. E. Cornell, et al., Hallmarks of reversible separation of living, unperturbed cell membranes into two liquid phases. *Biophys. J.* **113**, 2425–2432 (2017).

p. 87 "the yeast cells seem to use these domains to enable the digestion of stored fats"
- A. Y. Seo, P.-W. Lau, D. Feliciano, et al., AMPK and vacuole-associated Atg14p orchestrate μ-lipophagy for energy production and long-term survival under glucose starvation. *eLife* **6**, e21690 (2017).

p. 88 "This picture of cell membranes as two-dimensional fluids made possible by the self-organized lipid bilayer was cemented in the early 1970s"

• S. J. Singer, G. L. Nicolson, The fluid mosaic model of the structure of cell membranes. *Science* **175**, 720–731 (1972).

CHAPTER 6. PREDICTABLE RANDOMNESS

pp. 91–92 On the history of Brownian motion:
• R. M. Mazo, *Brownian Motion: Fluctuations, Dynamics, and Applications* (Clarendon Press, Oxford, UK, 2002).
• P. Hänggi, F. Marchesoni, 100 years of Brownian motion. *Chaos* **15**, 026101–026105 (2005).

pp. 92–95 A classic on the properties of random walks, and their importance in biology:
• H. C. Berg, *Random Walks in Biology* (Princeton University Press, Princeton, NJ, 1993).

p. 98 "My brain is relatively slow, but its neurons are much more interconnected than the transistors in my laptop's central processing unit"
• L. Luo, Why is the human brain so efficient? *Nautilus*, April 12, 2018. http://nautil.us/issue/59/connections/why-is-the-human-brain-so -efficient.

p. 99 "Finding a specific target like a DNA binding site is even more challenging. The average time required, it turns out, is roughly proportional to the cell size cubed"
• S. Redner, *A Guide to First-Passage Processes* (Cambridge University Press, Cambridge, UK, 2007).

p. 100 On motor proteins and transport in neurons:
• O. Yagensky, T. Kalantary Dehaghi, J. J. E. Chua, The roles of microtubule-based transport at presynaptic nerve terminals. *Front. Synaptic Neurosci.* **8**, 3 (2016).

p. 101 On bacterial motion and foraging:
• E. M. Purcell, "Life at low Reynolds number," *American Journal of Physics* **45**, 3–11 (1977).

CHAPTER 7. ASSEMBLING EMBRYOS

p. 105 "it was widely believed that this single cell held within it a *homunculus*, a miniature but fully formed human"
• C. Pinto-Correia, *The Ovary of Eve* (University of Chicago Press, Chicago, 1998).

p. 106 On Hans Driesch, including his statement that every embryonic cell "carries the totality of all primordia":

- S. F. Gilbert, *Developmental Biology* (Sinauer Associates, Sunderland, MA, sixth edition, 2000).

p. 107 "Christiane Nüsslein-Volhard and Eric Wieschaus discovered several genes that are important determinants of the body plan of the fruit fly"

- C. Nüsslein-Volhard, E. Wieschaus, Mutations affecting segment number and polarity in *Drosophila*. *Nature* **287**, 795–801 (1980).

p. 107 "naming one 'hedgehog' because mutations in it give rise to spiky fly larvae"

- D. R. Haskett, Hedgehog signaling pathway. *Embryo Project Encyclopedia* (2015). http://embryo.asu.edu/handle/10776/8685.

p. 108 The drawing of the fruit fly (*Drosophila melanogaster*) hedgehog protein is based on structure 2IBG in the Protein Data Bank:

- https://www.rcsb.org/structure/2IBG.
- J. S. McLellan, S. Yao, X. Zheng, et al., Structure of a heparin-dependent complex of hedgehog and ihog. *Proc. Natl. Acad. Sci.* **103**, 17208–17213 (2006).

p. 108 The drawing of the human sonic hedgehog protein is based on structure 3MXW in the Protein Data Bank:

- https://www.rcsb.org/structure/3MXW.
- H. R. Maun, X. Wen, A. Lingel, et al., Hedgehog pathway antagonist 5E1 binds hedgehog at the pseudo-active site. *J. Biol. Chem.* **285**, 26570–26580 (2010).

p. 110 "Transplanting tissue from the protein-producing region of one chick wing bud to the low-concentration side of another wing bud"

- M. Towers, J. Signolet, A. Sherman, et al., Insights into bird wing evolution and digit specification from polarizing region fate maps. *Nature Communications* **2**, 426 (2011).

p. 111 "Hedgehog gradients curate . . . the array of suckers on a cuttlefish arm"

- O. A. Tarazona, D. H. Lopez, L. A. Slota, M. J. Cohn, Evolution of limb development in cephalopod mollusks. *eLife* **8**, e43828 (2019).

p. 111 On hedgehog and cancer, see, for example:

- S. Kim, Y. Kim, J. Kong, et al., Epigenetic regulation of mammalian hedgehog signaling to the stroma determines the molecular subtype of bladder cancer. *eLife* **8**, e43024 (2019).

p. 111 "Morphogens were predicted and named by mathematician and computer science pioneer Alan Turing in 1952"

- M. Turing, The chemical basis of morphogenesis. *Philosophical Transactions of the Royal Society London, B: Biological Sciences* **237**, 37–72 (1952).

p. 114 "fluorescent glow provides a precise, quantifiable reporter of the what, where, and when of developmental activity"

- K. M. Forrest, E. R. Gavis, Live imaging of endogenous RNA reveals a diffusion and entrapment mechanism for nanos mRNA localization in *Drosophila*. *Current Biology* **13**, 1159–1168 (2003).
- T. Lucas, H. Tran, C.A.P. Romero, et al., 3 minutes to precisely measure morphogen concentration. *PLOS Genetics* **14**, e1007676 (2018).

pp. 114–115 "one can write down the equations of Brownian motion and genetic response functions and calculate patterns of protein abundance and gene activity"

- G. R. Ilsley, J. Fisher, R. Apweiler, et al., Cellular resolution models for even skipped regulation in the entire *Drosophila* embryo. *eLife* **2**, e00522 (2013).
- M. D. Petkova, G. Tkačik, W. Bialek, et al., Optimal decoding of cellular identities in a genetic network. *Cell* **176**, 844–855.e15 (2019).

p. 116 "William Bialek and colleagues at Princeton University connected morphogen measurements and information theory"

- J. O. Dubuis, G. Tkačik, E. F. Wieschaus, et al., Positional information, in bits. *Proc. Natl. Acad. Sci.* **110**, 16301–16308 (2013).

p. 117 On hair cells and lateral inhibition:

- M. Eddison, I. L. Roux, J. Lewis, Notch signaling in the development of the inner ear: Lessons from *Drosophila*. *Proc. Natl. Acad. Sci.* **97**, 11692–11699 (2000).

pp. 117–118 "Lateral inhibition . . . was first clearly demonstrated in a developing animal in the mid-1980s"

- C. Q. Doe, C. S. Goodman, Early events in insect neurogenesis: II. The role of cell interactions and cell lineage in the determination of neuronal precursor cells. *Developmental Biology* **111**, 206–219 (1985).

pp. 117–118 On the discovery of lateral inhibition in developmental patterning:

- K. Bussell, Milestone 3 (1937): Inhibit thy neighbour. *Nat. Rev. Neurosci.*, July 1, 2004. https://doi.org/10.1038/nrn1451.

pp. 117–118 On the Notch protein, its cleavage, and its use in cell signaling:

- W. R. Gordon, K. L. Arnett, S. C. Blacklow, The molecular logic of Notch signaling—a structural and biochemical perspective. *Journal of Cell Science* **121**, 3109–3119 (2008).

pp. 118–119 On Notch and lateral inhibition, more generally:

- M. Sjöqvist, E. R. Andersson, Do as I say, Not(ch) as I do: Lateral control of cell fate. *Developmental Biology* **447**, 58–70 (2019).

p. 119 On somite number in various animals and the early development of snakes:

- C. Gomez, E. M. Özbudak, J. Wunderlich, et al., Control of segment number in vertebrate embryos. *Nature* **454**, 335–339 (2008).

p. 120 "Jonathan Cooke and Erik Christopher Zeeman described an elegant biophysical strategy"

- J. Cooke, E. C. Zeeman, A clock and wavefront model for control of the number of repeated structures during animal morphogenesis. *J. Theor. Biol.* **58**, 455–476 (1976).

pp. 120–123 On the somite segmentation clock:

- I. Palmeirim, D. Henrique, D. Ish-Horowicz, O. Pourquié, Avian hairy gene expression identifies a molecular clock linked to vertebrate segmentation and somitogenesis. *Cell* **91**, 639–648 (1997).
- A. C. Oates, L. G. Morelli, S. Ares, Patterning embryos with oscillations: Structure, function and dynamics of the vertebrate segmentation clock. *Development* **139**, 625–639 (2012).

p. 123 "A joke already decades old." See, for example:

- What is the future of developmental biology? *Cell* **170**, 6–7 (2017).

CHAPTER 8. ORGANS BY DESIGN

p. 124 "you'll shed more than a ton of the cells lining your intestine"

- C. S. Potten, R. J. Morris, Epithelial stem cells in vivo. *J. Cell Sci.* **1988**, 45–62 (1988). (The stated value of about 10^{11} cells per day, at a mass of about 10^{-12} kg per cell, corresponds to about 3000 kg per lifetime, or about 6000 pounds.)

p. 125 "Dennis Discher and colleagues at the University of Pennsylvania grew stem cells of a type that can form neurons, muscle progenitors, or bone progenitors on gels of different stiffnesses"

- J. Engler, S. Sen, H. L. Sweeney, D. E. Discher, Matrix elasticity directs stem cell lineage specification. *Cell* **126**, 677–689 (2006).

pp. 125–126 On stem cells and their mechanical environment:

- J. Keung, S. Kumar, D. V. Schaffer, Presentation counts: Microenvironmental regulation of stem cells by biophysical and material cues. *Annual Review of Cell and Developmental Biology* **26**, 533–556 (2010).

p. 126 "channel-forming membrane proteins . . . whose configuration can be controlled by tension applied to the membrane"

- E. S. Haswell, R. Phillips, D. C. Rees, Mechanosensitive channels: What can they do and how do they do it? *Structure* **19**, 1356–1369 (2011).
- R. Peyronnet, D. Tran, T. Girault, J.-M. Frachisse, Mechanosensitive channels: Feeling tension in a world under pressure. *Front. Plant Sci.* **5**, 558 (2014).

pp. 127–128 "The groups . . . examined mice in which an expanding gel had been placed under the skin"

- M. Aragona, A. Sifrim, M. Malfait, et al., Mechanisms of stretch-mediated skin expansion at single-cell resolution. *Nature* **584**, 268–273 (2020).

p. 128 "fluid flow can induce stem cells to transform into the cell types that line blood vessels"

- K. Yamamoto, T. Sokabe, T. Watabe, et al., Fluid shear stress induces differentiation of Flk-1-positive embryonic stem cells into vascular endothelial cells in vitro. *American Journal of Physiology-Heart and Circulatory Physiology* **288**, H1915–H1924 (2005).
- H. Wang, G. M. Riha, S. Yan, et al., Shear stress induces endothelial differentiation from a murine embryonic mesenchymal progenitor cell line. *Arteriosclerosis, Thrombosis, and Vascular Biology* **25**, 1817–1823 (2005).

p. 129 On the history of cell culture, including Ross Harrison's experiments:

- H. Landecker, *Culturing Life: How Cells Became Technologies* (Harvard University Press, Cambridge, MA, 2010).

p. 129 "embryos of various species could be split apart into discrete cells that . . . could coalesce into aggregates that recapitulate some aspects of normal embryonic form." See, for example:

- M. S. Steinberg, Does differential adhesion govern self-assembly processes in histogenesis? Equilibrium configurations and the emergence of a hierarchy among populations of embryonic cells. *J. Exp. Zool.* **173**, 395–433 (1970).
- M. S. Steinberg, M. Takeichi, Experimental specification of cell sorting, tissue spreading, and specific spatial patterning by quantitative differences in cadherin expression. *Proc. Natl. Acad. Sci.* **91**, 206–209 (1994).

pp. 129–131 On the history of organoids:

- M. Simian, M. J. Bissell, Organoids: A historical perspective of thinking in three dimensions. *J. Cell Biol.* **216**, 31–40 (2017).

p. 130–131 On intestinal organoids:

- T. Sato, R. G. Vries, H. J. Snippert, et al., Single Lgr5 stem cells build crypt-villus structures in vitro without a mesenchymal niche. *Nature* **459**, 262–265 (2009).

p. 131 On organoids that form into a structure like the "optic cup":

- M. Eiraku, N. Takata, H. Ishibashi, M. Kawada, E. Sakakura, S. Okuda, K. Sekiguchi, T. Adachi, Y. Sasai, Self-organizing optic-cup morphogenesis in three-dimensional culture. *Nature* **472**, 51–56 (2011).

p. 131 "mouse stem cells could be grown into balls of connected neurons"

- M. Eiraku, K. Watanabe, M. Matsuo-Takasaki, et al., Self-organized formation of polarized cortical tissues from ESCs and its active manipulation by extrinsic signals. *Cell Stem Cell* **3**, 519–532 (2008).

p. 131 "the lab of Juergen Knoblich at the Austrian Academy of Science in Vienna built 'cerebral organoids'"

- M. A. Lancaster, M. Renner, C.-A. Martin, et al., Cerebral organoids model human brain development and microcephaly. *Nature* **501**, 373–379 (2013).

p. 131 "scientists and philosophers are already collaborating to map the ethical issues involved"

- J. Cepelewicz, An ethical future for brain organoids takes shape. *Quanta Magazine*, January 13, 2020. https://www.quantamagazine.org/an-ethical-future-for-brain-organoids-takes-shape-20200123/.

pp. 132–133 On the "lung on a chip":

- D. Huh, B. D. Matthews, A. Mammoto, et al., Reconstituting organ-level lung functions on a chip. *Science* **328**, 1662–8 (2010).

p. 133 On "body-on-a-chip" devices:

- C. W. McAleer, C. J. Long, D. Elbrecht, et al., Multi-organ system for the evaluation of efficacy and off-target toxicity of anticancer therapeutics. *Science Translational Medicine* **11**, eaav1386 (2019).
- C. D. Edington, W. L. K. Chen, E. Geishecker, et al., Interconnected microphysiological systems for quantitative biology and pharmacology studies. *Sci. Rep.* **8**, 1–18 (2018).

CHAPTER 9. THE ECOSYSTEM INSIDE YOU

p. 134 On the number of human and bacterial cells in the human body:

- R. Sender, S. Fuchs, R. Milo, Revised estimates for the number of human and bacteria cells in the body. *PLOS Biology* **14**, e1002533 (2016).

p. 135 "scooped a few hundred liters of water from the Sargasso Sea . . . and discovered a million previously unknown genes from hundreds of novel bacteria"

- J. C. Venter, K. Remington, J. F. Heidelberg, et al., Environmental genome shotgun sequencing of the Sargasso Sea. *Science* **304**, 66–74 (2004).

pp. 137–139 On the gut microbiota in general:

- J. Durack, S. V. Lynch, The gut microbiome: Relationships with disease and opportunities for therapy. *Journal of Experimental Medicine* **216**, 20–40 (2019).
- A. E. Douglas, *Fundamentals of Microbiome Science: How Microbes Shape Animal Biology* (Princeton University Press, Princeton, NJ, 2018).
- D. Haller, ed., *The Gut Microbiome in Health and Disease* (Springer, Cham, Switzerland, 2018).

p. 137 "There's also considerable, but not perfect, overlap between the set of species in you now and the set inside you a few months ago"

- B. H. Schlomann, R. Parthasarathy, Timescales of gut microbiome dynamics. *Current Opinion in Microbiology* **50**, 56–63 (2019).

p. 138 On the gut microbiome and neurological disorders:
- H. Tremlett, K. C. Bauer, S. Appel-Cresswell, et al., The gut microbiome in human neurological disease: A review. *Ann. Neurol.* **81**, 369–382 (2017).
- J. A. Griffiths, S. K. Mazmanian, Emerging evidence linking the gut microbiome to neurologic disorders. *Genome Medicine* **10**, 98 (2018).

p. 139 On fecal microbiota transplantation:
- L. Drew, Microbiota: Reseeding the gut. *Nature* **540**, S109–S112 (2016).
- E. van Nood, A. Vrieze, M. Nieuwdorp, et al., Duodenal infusion of donor feces for recurrent clostridium difficile. *New England Journal of Medicine* **368**, 407–415 (2013).
- R. J. Colman, D. T. Rubin, Fecal microbiota transplantation as therapy for inflammatory bowel disease: A systematic review and meta-analysis. *J. Crohns Colitis* **8**, 1569–1581 (2014).

p. 140 "compared the microbiome of fecal samples with that obtained by more invasive, direct sampling of the intestine"
- N. Zmora, G. Zilberman-Schapira, J. Suez, et al., Personalized gut mucosal colonization resistance to empiric probiotics is associated with unique host and microbiome features. *Cell* **174**, 1388–1405.e21 (2018).

p. 141 "germ-free animals exhibit a wide range of abnormalities"
- J. L. Round, S. K. Mazmanian, The gut microbiota shapes intestinal immune responses during health and disease. *Nat. Rev. Immunol.* **9**, 313–323 (2009).
- L. V. Hooper, D. R. Littman, A. J. Macpherson, Interactions between the microbiota and the immune system. *Science* **336**, 1268–1273 (2012).
- T. A. Jones, K. Guillemin, Racing to stay put: How resident microbiota stimulate intestinal epithelial cell proliferation. *Curr. Pathobiol. Rep.* **6**, 23–28 (2018).

p. 141 "germ-free fish larvae have a paucity of the insulin-producing beta cells of the pancreas"
- J. H. Hill, E. A. Franzosa, C. Huttenhower, K. Guillemin, A conserved bacterial protein induces pancreatic beta cell expansion during zebrafish development. *eLife* **5**, e20145 (2016).

p. 141 "Microbiome-assisted development has been observed for several other organs and tissues, for example, bone growth in infant mice"
- M. Schwarzer, K. Makki, G. Storelli, et al., Lactobacillus plantarum strain maintains growth of infant mice during chronic undernutrition. *Science* **351**, 854–857 (2016).

p. 142 "Links between human obesity and the composition of the gut microbiome"

- M. M. Finucane, T. J. Sharpton, T. J. Laurent, K. S. Pollard, A taxonomic signature of obesity in the microbiome? Getting to the guts of the matter. *PLoS ONE* **9**, e84689 (2014).
- M. A. Sze, P. D. Schloss, Looking for a signal in the noise: Revisiting obesity and the microbiome. *mBio.* **7**, e01018–16 (2016).

p. 144 On the prevalence and treatment of cholera:

- See articles on the websites of the World Health Organization (https://www.who.int/news-room/fact-sheets/detail/cholera) and the US Centers for Disease Control and Prevention (https://www.cdc.gov/cholera/treatment/index.html).

p. 145 On the type VI secretion system in *Vibrio cholerae* and other bacteria:

- B. Russell, S. B. Peterson, J. D. Mougous, Type VI secretion system effectors: Poisons with a purpose. *Nat. Rev. Micro.* **12**, 137–148 (2014).

pp. 145–147 On my lab's experiments on *Vibrio cholerae* and its type VI secretion system:

- S. L. Logan, J. Thomas, J. Yan, R. P. Baker, D. S. Shields, J. B. Xavier, B. K. Hammer, R. Parthasarathy, The *Vibrio cholerae* type VI secretion system can modulate host intestinal mechanics to displace gut bacterial symbionts. *Proc. Natl. Acad. Sci.* **115**, E3779–E3787 (2018).

p. 149 "we saw large immune responses when the fish was colonized by the normal, motile bacteria"

- T. J. Wiles, B. H. Schlomann, E. S. Wall, et al., Swimming motility of a gut bacterial symbiont promotes resistance to intestinal expulsion and enhances inflammation. *PLOS Biology* **18**, e3000661 (2020).

p. 149 "microbes with memories could record intestinal conditions"

- J. W. Kotula, S. J. Kerns, L. A. Shaket, et al., Programmable bacteria detect and record an environmental signal in the mammalian gut. *Proc. Natl. Acad. Sci.* **111**, 4838–4843 (2014).

p. 150 "devices that mimic aspects of the pulsatile flow of the human gut"

- J. Cremer, I. Segota, C. Yang, et al., Effect of flow and peristaltic mixing on bacterial growth in a gut-like channel. *Proc. Natl. Acad. Sci.* **113**, 11414–11419 (2016).

p. 150 "stretchable gut-on-a-chip devices"

- W. Shin, C. D. Hinojosa, D. E. Ingber, H. J. Kim, Human intestinal morphogenesis Controlled by Transepithelial Morphogen Gradient and Flow-Dependent Physical Cues in a microengineered gut-on-a-chip. *iScience* **15**, 391–406 (2019).

p. 151 "Alvaro Sanchez and colleagues . . . put together hundreds of microbial communities derived from samples of soil and plant leaves"

- J. E. Goldford, N. Lu, D. Bajić, et al., Emergent simplicity in microbial community assembly. *Science* **361**, 469–474 (2018).

p. 152 "ecologist Robert May, who in classic and highly influential work in the 1970s"

- R. M. May, *Stability and Complexity in Model Ecosystems* (Princeton University Press, Princeton, NJ, reprint edition, 2001).
- R. M. May, Will a large complex system be stable? *Nature* **238**, 413–414 (1972).

p. 152 "coexistence can emerge beyond the destabilizing point theorized by May"

- W. Cui, R. Marsland, P. Mehta, Diverse communities behave like typical random ecosystems. bioRxiv.org, 596551 (2019). https://doi.org/10.1101 /596551.
- R. Marsland III, W. Cui, J. Goldford, et al., Available energy fluxes drive a transition in the diversity, stability, and functional structure of microbial communities. *PLOS Computational Biology* **15**, e1006793 (2019).

p. 152 "a mathematical description of this sort of constrained resource usage finds a surprisingly large set of parameters for which coexistence occurs"

- A. Posfai, T. Taillefumier, N. S. Wingreen, Metabolic trade-offs promote diversity in a model ecosystem. *Phys. Rev. Lett.* **118**, 028103 (2017).

CHAPTER 10. A SENSE OF SCALE

p. 155 "Large organisms don't simply look like blown-up versions of small ones." Several fascinating works expand on the themes of this chapter:

- S. Vogel, *Life's Devices: The Physical World of Animals and Plants* (Princeton University Press, Princeton, NJ, 1988).
- K. Schmidt-Nielsen, *How Animals Work* (Cambridge University Press, Cambridge, UK, 1972).
- J.B.S. Haldane, *On Being the Right Size and Other Essays* (Oxford University Press, Oxford, UK, 1985).

p. 158 "fluid dynamics pioneer Osborne Reynolds . . . became in 1868 the second ever 'professor of engineering' in England"

- J. J. O'Connor, E. F. Robertson, Osborne Reynolds—biography. *Maths History* (2003). https://mathshistory.st-andrews.ac.uk/Biographies/Reynolds/.

pp. 159–161 On the importance of the Reynolds number in biology:

- S. Vogel, *Life in Moving Fluids: The Physical Biology of Flow* (Princeton University Press, Princeton, NJ, second edition, 1996).

p. 160 "A classic video of this effect by fluid dynamicist G. I. Taylor is available online"

- *Low Reynolds Number Flow* at the National Committee for Fluid Mechanics Films site, http://web.mit.edu/hml/ncfmf.html; also at https://www.youtube.com/watch?v=51-6QCJTAjU (start at 13:40).

p. 166 "a plot of mass versus leg length for several different cockroaches"
- Henry D. Prange, "The scaling and mechanics of arthropod exoskeletons," in *Scale Effects in Animal Locomotion*, ed. T. J. Pedley (Academic Press, London, 1997), 169–171.

p. 167 "a 1963 paper on lung physiology"
- S. M. Tenney, J. E. Remmers, Comparative quantitative morphology of the mammalian lung: Diffusing area. *Nature* **197**, 54–56 (1963).

p. 168–169 "Thomas McMahon and John Tyler Bonner . . . plot the diameter versus the end-to-end length of the humerus, a leg bone, of lots of bovids"
- T. A. McMahon, J. T. Bonner, *On Size and Life* (Scientific American Library, New York, 1983).

p. 171 "an elephant named Tusko . . . lived a mostly miserable life in early twentieth-century circuses"
- M. Cooper, The elephant in the room. *Oregon Quarterly* (Spring 2014), 34–38.
- C. Lynn, The time Tusko the elephant was abandoned at the Oregon State Fair. *Salem Statesman Journal*, August 25, 2017.

p. 171 "If they did, bones would make up about three-quarters of their body mass"
- S. Vogel, *Life's Devices: The Physical World of Animals and Plants* (Princeton University Press, Princeton, NJ, 1988).

p. 172 "Galileo, in fact, wrote about bone scaling in his 1638 book, *Two New Sciences*"
- Galileo Galilei, *Dialogues Concerning Two New Sciences*, trans. Stillman Drake (University of Wisconsin Press, Madison, 1974).

CHAPTER 11. LIFE AT THE SURFACE

pp. 173–175 On respiration in ants and other insects:
- C. Gillott, ed., *Entomology* (Springer, Dordrecht, Netherlands, 2005), 469–486.
- M. J. Klowden, *Physiological Systems in Insects* (Academic Press, Amsterdam, 2010).

p. 175 "we find in the fossil record abundant evidence of giant insects"
- Ker Than, Why giant bugs once roamed the earth. *National Geographic*, August 9, 2011. https://www.nationalgeographic.com/news/2011/8/110808-ancient-insects-bugs-giants-oxygen-animals-science/.

p. 179 On fire ants and their strategies for staying atop water:
- N. J. Mlot, C. A. Tovey, D. L. Hu, Fire ants self-assemble into waterproof rafts to survive floods. *Proc. Natl. Acad. Sci.* **108**, 7669–7673 (2011).

p. 180 "infant respiratory distress syndrome, or IRDS, the leading cause of death among premature infants"
- L. K. Altman, A Kennedy baby's life and death. *New York Times*, July 30, 2013. https://www.nytimes.com/2013/07/30/health/a-kennedy-babys -life-and-death.html.
- D. Schraufnagel, *Breathing in America: Diseases, Progress, and Hope* (American Thoracic Society, New York, first edition, 2010).

pp. 180–181 On surface tension in the lungs and the role of pulmonary surfactant:
- J. A. Clements, Surface tension in the lungs. *Scientific American* **207**, 120– 130 (1962).
- J. A. Clements, Lung surfactant: A personal perspective. *Annual Review of Physiology* **59**, 1–21 (1997).
- L. G. Dobbs, Pulmonary surfactant. *Annu. Rev. Med.* **40**, 431–446 (1989).

CHAPTER 12. MYSTERIES OF SIZE AND SHAPE

p. 183 "If you're curious about these examples, I've listed some readings in the references"
- T. A. McMahon, J. T. Bonner, *On Size and Life* (Scientific American Library, New York, 1983).
- M. Gazzola, M. Argentina, L. Mahadevan, Scaling macroscopic aquatic locomotion. *Nat. Phys.* **10**, 758–761 (2014).
- J. Baumgart, B. M. Friedrich, Fluid dynamics: Swimming across scales. *Nat. Phys.* **10**, 711–712 (2014).
- P. Willmer, G. Stone, I. Johnston, *Environmental Physiology of Animals* (Wiley-Blackwell, Malden, MA, 2004).
- R. Bale, M. Hao, A.P.S. Bhalla, N. A. Patankar, Energy efficiency and allometry of movement of swimming and flying animals. *Proc. Natl. Acad. Sci.* **111**, 7517–7521 (2014).
- T. L. Hedrick, B. Cheng, X. Deng, Wingbeat time and the scaling of passive rotational damping in flapping flight. *Science* **324**, 252–255 (2009).

p. 184 "an evocative description from biochemist and writer Nick Lane, 'an elephant-sized pile of mice would consume twenty times more food and oxygen every minute than the elephant does itself'"
- N. Lane, *Power, Sex, Suicide: Mitochondria and the Meaning of Life* (Oxford University Press, Oxford, UK, 2005).

p. 185 "Max Kleiber considered a range of animals from doves to cattle, plotting their metabolic rates and body masses"
- M. Kleiber, Body size and metabolism. *Hilgardia* **6**, 315–353 (1932).

p. 185 "from a 2003 dataset of over 600 mammals"
- C. R. White, R. S. Seymour, Mammalian basal metabolic rate is proportional to body mass$^{2/3}$. *Proc. Natl. Acad. Sci.* **100**, 4046–4049 (2003).

p. 186 "Max Rubner examined the energy consumption of seven dog breeds"
- M. Rubner, Ueber den einfluss der korpergrosse auf stoffund kaftwechsel. *Zeitschrift fur Biologie* **19**, 535–562 (1883).

p. 186 "Oxygen consumption in guinea pigs shows the same behavior"
- T. A. McMahon, J. T. Bonner, *On Size and Life* (Scientific American Library, New York, 1983), 56.

p. 187 "Peter Dodds, Dan Rothman, and Joshua Weitz at the Massachusetts Institute of Technology evaluated several datasets"
- P. S. Dodds, D. H. Rothman, J. S. Weitz, Re-examination of the "3/4-law" of metabolism. *Journal of Theoretical Biology* **209**, 9–27 (2001).

p. 187 "Considering birds alone, Peter Bennett and Paul H. Harvey . . . found a scaling exponent of 0.67"
- P. M. Bennett, P. H. Harvey, Active and resting metabolism in birds: Allometry, phylogeny and ecology. *Journal of Zoology* **213**, 327–344 (1987).

p. 189 "Clouds are not spheres, mountains are not cones, coastlines are not circles"
- Benoit Mandelbrot, *The Fractal Geometry of Nature* (W. H. Freeman, San Francisco, 1982).

p. 189 "West, Brown, and Enquist proposed a creative explanation for metabolic scaling"
- G. B. West, J. H. Brown, B. J. Enquist, A general model for the origin of allometric scaling laws in biology. *Science* **276**, 122–126 (1997).

p. 190 "several researchers found subtle but important flaws in West and colleagues' proof"
- P. S. Dodds, D. H. Rothman, J. S. Weitz, Re-examination of the "3/4-law" of metabolism. *Journal of Theoretical Biology* **209**, 9–27 (2001).
- R. S. Etienne, M.E.F. Apol, H. Olff, Demystifying the West, Brown & Enquist model of the allometry of metabolism. *Functional Ecology* **20**, 394–399 (2006).

pp. 190–191 "Jayanth Banavar . . . Amos Maritan, and Andrea Rinaldo developed a mathematical model that doesn't require self-similarity"
- J. R. Banavar, A. Maritan, A. Rinaldo, Size and form in efficient transportation networks. *Nature* **399**, 130–132 (1999).

p. 192 "This perspective has been nicely described by biochemist and author Nick Lane"

- N. Lane, *Power, Sex, Suicide: Mitochondria and the Meaning of Life* (Oxford University Press, Oxford, UK, 2005). (See chapter 9, "The Power Laws of Biology.")

p. 192 "This perspective . . . was precisely formulated by Peter Hochachka's group"
- C.-A. Darveau, R. K. Suarez, R. D. Andrews, P. W. Hochachka, Allometric cascade as a unifying principle of body mass effects on metabolism. *Nature.* **417**, 166–170 (2002).

p. 192 "Brown, West, and others have argued for a 'metabolic theory of ecology'"
- J. H. Brown, J. F. Gillooly, A. P. Allen, V. M. Savage, G. B. West, Toward a metabolic theory of ecology. *Ecology* **85**, 1771–1789 (2004).

p. 193 "Geoffrey West and others . . . have argued that cities, like living organisms, are governed by the properties of networks"
- L.M.A. Bettencourt, J. Lobo, D. Helbing, et al., Growth, innovation, scaling, and the pace of life in cities. *Proc. Natl. Acad. Sci.* **104**, 7301–7306 (2007).
- L.M.A. Bettencourt, The origins of scaling in cities. *Science* **340**, 1438–1441 (2013).

pp. 193–194 "The data are, however, noisy, and their trends are even more contentious than those of animal metabolism"
- C. R. Shalizi, Scaling and hierarchy in urban economies. arXiv.org, 1102.4101 [physics, stat.] (2011). http://arxiv.org/abs/1102.4101.
- E. Arcaute, E. Hatna, P. Ferguson, H. Youn, A. Johansson, M. Batty, Constructing cities, deconstructing scaling laws. *Journal of the Royal Society Interface* **12**, 20140745 (2015).

p. 194 "flatworms that grow or shrink dramatically depending on the availability of food"
- A. Thommen, S. Werner, O. Frank, et al., Body size-dependent energy storage causes Kleiber's law scaling of the metabolic rate in planarians. *eLife* **8**, e38187 (2019).

CHAPTER 13. HOW WE READ DNA

p. 201 "Ray Wu and Dale Kaiser managed to decipher 12 of the nucleotides of a viral genome"
- R. Wu, A. D. Kaiser, Structure and base sequence in the cohesive ends of bacteriophage lambda DNA. *Journal of Molecular Biology* **35**, 523–537 (1968).

pp. 201–215 On the history of DNA sequencing:
- J. Shendure, S. Balasubramanian, G. M. Church, W. Gilbert, J. Rogers, J. A. Schloss, R. H. Waterston, DNA sequencing at 40: Past, present and future. *Nature* **550**, 345–353 (2017).

- J. M. Heather, B. Chain, The sequence of sequencers: The history of sequencing DNA. *Genomics* **107**, 1–8 (2016).

pp. 203–204 On DNA in electric fields:

- M. Muthukumar, Theory of electrophoretic mobility of polyelectrolyte chains. *Macromolecular Theory and Simulations* **3**, 61–71 (1994).
- J.-L. Viovy, Electrophoresis of DNA and other polyelectrolytes: Physical mechanisms. *Rev. Mod. Phys.* **72**, 813–872 (2000).

pp. 205–207 On the initial sequencing of the human genome:

- T. Carvalho, T. Zhu, The Human genome project (1990–2003). *Embryo Project Encyclopedia* (2014). http://embryo.asu.edu/handle/10776/7829.
- International Human Genome Sequencing Consortium, Initial sequencing and analysis of the human genome. *Nature* **409**, 860–921 (2001).
- J. C. Venter et al., The sequence of the human genome. *Science* **291**, 1304–1351 (2001).

p. 206 On the completion of the Human Genome Project:

- E. Pennisi, Reaching their goal early, sequencing labs celebrate. *Science* **300**, 409 (2003).
- Genome.gov, Human Genome Project FAQ. https://www.genome.gov/human-genome-project/Completion-FAQ.

p. 207 On second-generation DNA sequencing methods:

- E. R. Mardis, Next-generation DNA sequencing methods. *Annual Review of Genomics and Human Genetics* **9**, 387–402 (2008).
- M. L. Metzker, Sequencing technologies—the next generation. *Nature Reviews Genetics* **11**, 31–46 (2010).

pp. 207–210 On pyrosequencing:

- M. Margulies, M. Egholm, W. E. Altman, et al., Genome sequencing in microfabricated high-density picolitre reactors. *Nature* **437**, 376–380 (2005).
- P. Nyren, B. Pettersson, M. Uhlen, Solid phase DNA minisequencing by an enzymatic luminometric inorganic pyrophosphate detection assay. *Analytical Biochemistry* **208**, 171–175 (1993).

p. 210 "For about $500,000, you could buy a machine to perform pyrosequencing"

- K. Davies, *The $1,000 Genome: The Revolution in DNA Sequencing and the New Era of Personalized Medicine* (Free Press, New York, 2010).

p. 211 On DNA sequencing using field effect transistors (known as ion semiconductor sequencing):

- E. Pennisi, Semiconductors inspire new sequencing technologies. *Science* **327**, 1190 (2010).

- J. M. Rothberg, W. Hinz, T. M. Rearick, et al., An integrated semiconductor device enabling non-optical genome sequencing. *Nature* **475**, 348–352 (2011).

p. 211 "Ion Torrent introduced its Personal Genome Machine in 2010"

- M. Herper, Gene machine. *Forbes*, December 30, 2010. https://www.forbes.com/forbes/2011/0117/features-jonathan-rothberg-medicine-tech-gene-machine.html#12c8a7ed2711.

pp. 211–212 On Illumina sequencing:

- YouTube user Draven1983101, Illumina Solexa sequencing (2010). https://www.youtube.com/watch?v=77r5p8IBwJk.

p. 212 On Pacific Biosciences' DNA sequencing technique:

- J. Eid, A. Fehr, J. Gray, et al., Real-time DNA sequencing from single polymerase molecules. *Science* **323**, 133–138 (2009).

p. 213–215 On nanopore sequencing:

- H. Bayley, Nanopore sequencing: From imagination to reality. *Clin. Chem.* **61**, 25–31 (2015).
- D. Deamer, M. Akeson, D. Branton, Three decades of nanopore sequencing. *Nat. Biotechnol.* **34**, 518–524 (2016).

p. 214 The drawing of the nanopore channel protein is based on structure 3X2R in the Protein Data Bank:

- https://www.rcsb.org/structure/3X2R.
- B. Cao, Y. Zhao, Y. Kou, et al., Structure of the nonameric bacterial amyloid secretion channel. *Proc. Natl. Acad. Sci.* **111**, E5439–E5444 (2014).

p. 214 The drawing of the polymerase attached to the nanopore channel protein is based on structure 3BDP in the Protein Data Bank:

- https://www.rcsb.org/structure/3BDP.
- J. R. Kiefer, C. Mao, J. C. Braman, L. S. Beese, Visualizing DNA replication in a catalytically active Bacillus DNA polymerase crystal. *Nature* **391**, 304–307 (1998).

p. 216 "The graph [of genome sequencing costs] is amazing." The data are from a database maintained by the US National Institutes of Health:

- K. A. Wetterstrand, DNA sequencing costs: Data from the NHGRI genome sequencing program (GSP). www.genome.gov/sequencingcostsdata (accessed February 6, 2020).

p. 217 "public funding for basic research as well as specific programs targeting sequencing innovations." See, for example:

- https://www.nih.gov/news-events/news-releases/nhgri-funds-development-third-generation-dna-sequencing-technologies.

p. 218 On the discovery of reverse transcriptase:

- J. M. Coffin, H. Fan, The discovery of reverse transcriptase. *Annual Review of Virology* **3**, 29–51 (2016).

p. 218 On single-cell RNA sequencing:
- G. X. Y. Zheng, J. M. Terry, P. Belgrader, et al., Massively parallel digital transcriptional profiling of single cells. *Nature Communications* **8**, 14049 (2017).
- J. Shendure, E. Lieberman Aiden, The expanding scope of DNA sequencing. *Nat. Biotechnol.* **30**, 1084–1094 (2012).
- D. Kotliar, A. Veres, M. A. Nagy, et al., Identifying gene expression programs of cell-type identity and cellular activity with single-cell RNA-Seq. *eLife* **8**, e43803 (2019).
- D. E. Wagner, C. Weinreb, Z. M. Collins, et al., Single-cell mapping of gene expression landscapes and lineage in the zebrafish embryo. *Science* **360**, 981–987 (2018).

CHAPTER 14. GENETIC COMBINATIONS

p. 222 "The average Frenchman born in 1800 was 5 feet, 5 inches tall" and "Our modern Frenchman has about twice the calories per day at his disposal"
- M. Roser, C. Appel, H. Ritchie, Human height. *Our World in Data* (2013). https://ourworldindata.org/human-height.
- M. Roser, H. Ritchie, Food supply. *Our World in Data* (2013). https://ourworldindata.org/food-supply.

p. 223 On the heritability of height, nutrition, and whole genome studies:
- C.-Q. Lai, How much of human height is genetic and how much is due to nutrition? *Scientific American*, December 11, 2006. https://www.scientificamerican.com/article/how-much-of-human-height/.

pp. 225–227 On predicting height from SNP data:
- L. Lello, S. G. Avery, L. Tellier, et al., Accurate genomic prediction of human height. *Genetics* **210**, 477–497 (2018).

p. 227 "the information encoded by DNA does, in fact, explain 80% of the variation in human height"
- P. Wainschtein, D. P. Jain, L. Yengo, et al., Recovery of trait heritability from whole genome sequence data. bioRxiv.org, 588020 (2019).
- L. Geddes, Genetic study homes in on height's heritability mystery. *Nature* **568**, 444–445 (2019).

p. 227 On Norman Borlaug and wheat:
- L. F. Hesser, *The Man Who Fed the World: Nobel Peace Prize Laureate Norman Borlaug and His Battle to End World Hunger* (Durban House Publishing, Dallas, TX, 2006).

- G. Easterbrook, Forgotten benefactor of humanity. *The Atlantic* (January 1997) https://www.theatlantic.com/magazine/archive/1997/01/forgotten-benefactor-of-humanity/306101/.

p. 227 "Contemporary chickens raised for consumption in North America weigh four times as much as their ancestors of the 1950s"
- M. J. Zuidhof, B. L. Schneider, V. L. Carney, et al., Growth, efficiency, and yield of commercial broilers from 1957, 1978, and 2005. *Poultry Science* **93**, 2970–2982 (2014).

p. 228 "In 2019, the US dairy database, for example, contained genotypes from three million cows"
- Council on Dairy Cattle Breeding, Annual reports. https://www.uscdcb .com/whats-new/reports/ (accessed April 1, 2020).

p. 229 On watermelons:
- M. Strauss, The 5,000-year secret history of the watermelon. *National Geographic*, August 21, 2015. https://www.nationalgeographic.com/news /2015/08/150821-watermelon-fruit-history-agriculture/.
- M. Jayakodi, M. Schreiber, M. Mascher, Sweet genes in melon and watermelon. *Nature Genetics* **51**, 1572–1573 (2019).
- S. Guo, S. Zhao, H. Sun, et al., Resequencing of 414 cultivated and wild watermelon accessions identifies selection for fruit quality traits. *Nature Genetics* **51**, 1616–1623 (2019).

pp. 229–230 On breast cancer risks:
- A. Antoniou, P.D.P. Pharoah, S. Narod, et al., Average risks of breast and ovarian cancer associated with BRCA1 or BRCA2 mutations detected in case series unselected for family history: A combined analysis of 22 studies. *American Journal of Human Genetics* **72**, 1117–1130 (2003).

p. 231 "In 1969, Robert Edwards, Barry Bavister, and Patrick Steptoe announced that they had successfully fertilized human egg cells with human sperm cells in vitro"
- R. G. Edwards, B. D. Bavister, P. C. Steptoe, Early stages of fertilization in vitro of human oocytes matured in vitro. *Nature* **221**, 632–635 (1969).

pp. 231–232 "The intervening decade saw a great deal of discussion." For a much more thorough treatment of the scientific, historical, and social context of in vitro fertilization, see
- P. Ball, *Unnatural: The Heretical Idea of Making People* (Bodley Head, London, first edition, 2011).

p. 232 "A 1969 *Life* magazine cover"
- *Life*, June 13, 1969.

p. 232 "a poll of Americans in the same issue"
- L. Harris, The *Life* poll. *Life* **66** (23), 52–55 (June 13, 1969).

p. 232 "By 1978, however, according to a Gallup poll also conducted in the United States"

- H. M. Kiefer, Gallup brain: The birth of in vitro fertilization. Gallup.com, August 5, 2003. https://news.gallup.com/poll/8983/Gallup-Brain-Birth -Vitro-Fertilization.aspx.

p. 232 "There have been in total about eight million babies born worldwide through *in vitro* fertilization as of 2018"

- European Society of Human Reproduction and Embryology. ScienceDaily (website), July 3, 2018. https://www.sciencedaily.com/releases/2018/07 /180703084127.htm.

p. 232 On the fraction of babies born in Denmark through in vitro fertilization:

- Danish Health and Medicines Authority, 2017 assisted reproduction report. https://sundhedsdatastyrelsen.dk/da/tal-og-analyser/analyser -og-rapporter/andre-analyser-og-rapporter/assisteret-reproduktion (accessed October 26, 2020).

p. 233 On embryo biopsy methods:

- D. Cimadomo, A. Capalbo, F. M. Ubaldi, et al., The impact of biopsy on human embryo developmental potential during preimplantation genetic diagnosis. *BioMed Research International* (2016), e7193075.
- H. J. Stern, Preimplantation genetic diagnosis: Prenatal testing for embryos finally achieving its potential. *Journal of Clinical Medicine* 3, 280– 309 (2014).

p. 235 "one might expect embryo selection to give a height advantage of about 1 inch"

- E. Karavani, O. Zuk, D. Zeevi, et al., Screening human embryos for polygenic traits has limited utility. *Cell* 179, 1424–1435.e8 (2019).

p. 236 "About 30,000 forced sterilizations had been performed . . . by 1939"

- M. Wills, When forced sterilization was legal in the U.S. *JSTOR Daily*, August 3, 2017. https://daily.jstor.org/when-forced-sterilization-was -legal-in-the-u-s/.

p. 236 On eugenics, especially in the United States:

- A. DenHoed, The forgotten lessons of the American eugenics movement. *New Yorker,* April 27, 2016. https://www.newyorker.com/books/page -turner/the-forgotten-lessons-of-the-american-eugenics-movement.
- C. Zimmer, *She Has Her Mother's Laugh: The Powers, Perversions, and Potential of Heredity* (Dutton, New York, first edition, 2018).

p. 237 "Signatures of 'assortative mating,' in which people of similar educational or socioeconomic backgrounds pair up, are evident"

- M. R. Robinson, A. Kleinman, M. Graff, et al., Genetic evidence of assortative mating in humans. *Nature Human Behaviour* 1, 1–13 (2017).

CHAPTER 15. HOW WE WRITE DNA

p. 240 "insulin from pigs and cattle, the purification of which is difficult and expensive"

- D. Wendt, Two tons of pig parts: Making insulin in the 1920s. *O Say Can You See?* (blog for the National Museum of American History), November 1, 2013. https://americanhistory.si.edu/blog/2013/11/two-tons -of-pig-parts-making-insulin-in-the-1920s.html.
- Genentech (website), Cloning insulin, April 7, 2016. https://www.gene .com/stories/cloning-insulin.

pp. 241–243 A simple, illustrated explanation of bacterial transformation can be found at Khan Academy:

- https://www.khanacademy.org/science/biology/biotech-dna-techno logy/dna-cloning-tutorial/a/bacterial-transformation-selection.

pp. 241–243 More detailed bacterial transformation protocols that give a sense of what these procedures are like in practice can be found at the websites of vendors of supplies and chemicals. See, for example:

- https://www.thermofisher.com/us/en/home/life-science/cloning /cloning-learning-center/invitrogen-school-of-molecular-biology /molecular-cloning/transformation.html.

p. 243 "Working together, the labs of Stanley Cohen at Stanford University and Herbert Boyer at the University of California at San Francisco announced the first successful bacterial transformation"

- S. N. Cohen, A.C.Y. Chang, H. W. Boyer, R. B. Helling, Construction of biologically functional bacterial plasmids in vitro. *Proc. Natl. Acad. Sci.* **70**, 3240–3244 (1973).

pp. 243–244 On Genetech and the bioengineering of insulin-producing bacteria:

- S. S. Hughes, *Genentech: The Beginnings of Biotech* (University of Chicago Press, Chicago, reprint edition, 2013).
- S. Mukherjee, *The Gene: An Intimate History* (Scribner, New York, 2016).

p. 243 "As one of the councillors said later"

- B. J. Culliton, Recombinant DNA: Cambridge City Council votes moratorium. *Science* **193**, 300–301 (1976).

p. 244 "These days insulin itself is mostly produced from engineered yeast rather than bacteria"

- J. Nielsen, Production of biopharmaceutical proteins by yeast. *Bioengineered* **4**, 207–211 (2013).

p. 245 "DNA carefully injected into the nascent embryo can find itself inserted into the genome"

- R. Behringer, M. Gertsenstein, K. Nagy, A. Nagy, *Manipulating the Mouse Embryo: A Laboratory Manual* (Cold Spring Harbor Laboratory Press, Cold Spring Harbor, NY, fourth edition, 2013).

pp. 245–246 On agrobacterium and DNA insertion into plant genomes:

- T. Tzfira, J. Li, B. Lacroix, V. Citovsky, Agrobacterium T-DNA integration: Molecules and models. *Trends in Genetics* **20**, 375–383 (2004).

p. 246 "Inadequate vitamin A causes blindness in 250,000 to 500,000 children annually"

- World Health Organization, Micronutrient deficiencies: Vitamin A deficiency. https://www.who.int/nutrition/topics/vad/en/ (accessed September 15, 2020).

p. 246 "Researchers . . . developed 'golden rice'"

- X. Ye, S. Al-Babili, A. Klöti, et al., Engineering the provitamin A (β-carotene) biosynthetic pathway into (carotenoid-free) rice endosperm. *Science* **287**, 303–305 (2000).

pp. 246–247 Abundant information about golden rice, including details of the genetics and safety tests, is at

- Golden Rice Humanitarian Board, Golden Rice Project. http://www.goldenrice.org/.

p. 246 "Clinical studies showed that the strikingly yellow-orange rice is safe and effective"

- G. Tang, J. Qin, G. G. Dolnikowski, et al., Golden rice is an effective source of vitamin A. *Am. J. Clin. Nutr.* **89**, 1776–1783 (2009).

p. 246 "Golden Rice could probably supply 50% of the Recommended Dietary Allowance (RDA) of vitamin A"

- American Society of Nutrition, Researchers determine that golden rice is an effective source of vitamin A. http://www.goldenrice.org/PDFs/ASNonGR.pdf (May 15, 2009).

p. 247 On the history and struggles of golden rice:

- E. Regis, *Golden Rice: The Imperiled Birth of a GMO Superfood* (Johns Hopkins University Press, Baltimore, 2019).
- E. Regis, The true story of the genetically modified superfood that almost saved millions. *Foreign Policy*, October 17, 2019. https://foreignpolicy.com/2019/10/17/golden-rice-genetically-modified-superfood-almost-saved-millions/.

p. 247 "Bangladesh . . . became in 2019 the first country to approve the planting of golden rice seeds"

- E. Stokstad, After 20 years, golden rice nears approval. *Science* **366**, 934–934 (2019).

p. 247 "the Philippines acknowledged the safety of golden rice, setting the stage for its planting"

- J. Conrow, Philippine agency rules golden rice is safe. Cornell Alliance for Science December 18, 2019. https://allianceforscience.cornell.edu /blog/2019/12/philippine-agency-rules-golden-rice-is-safe/.

p. 248 On zinc finger nucleases and transcription activator-like effector nucleases:

- T. Gaj, C. A. Gersbach, C. F. Barbas, ZFN, TALEN, and CRISPR/Cas-based methods for genome engineering. *Trends in Biotechnology* **31**, 397–405 (2013).

p. 248 "In 1987, a Japanese team"

- Y. Ishino, H. Shinagawa, K. Makino, et al., Nucleotide sequence of the iap gene, responsible for alkaline phosphatase isozyme conversion in *Escherichia coli*, and identification of the gene product. *Journal of Bacteriology* **169**, 5429–5433 (1987).

p. 248 "Francisco Mojica and colleageus in Spain discovered similar repeating DNA patterns . . ."

- F. J. M. Mojica, G. Juez, F. Rodriguez-Valera, Transcription at different salinities of Haloferax mediterranei sequences adjacent to partially modified PstI sites. *Molecular Microbiology*. **9**, 613–621 (1993).

pp. 248–249 On the early history of CRISPR:

- M. Campbell, Francis Mojica: The modest microbiologist who discovered and named CRISPR. Genomics Research from Technology Networks (website), October 14, 2019. https://www.technologynetworks.com /genomics/articles/francis-mojica-the-modest-microbiologist-who -discovered-and-named-crispr-325093.
- P. D. Hsu, E. S. Lander, F. Zhang, Development and applications of CRISPR-Cas9 for genome engineering. *Cell* **157**, 1262–1278 (2014).
- Y. Ishino, M. Krupovic, P. Forterre, History of CRISPR-Cas from encounter with a mysterious repeated sequence to genome editing technology. *Journal of Bacteriology* **200**, e00580-17 (2018). doi:10.1128/ JB.00580-17.
- R. Barrangou, J. van der Oost, eds., *CRISPR-Cas Systems: RNA-Mediated Adaptive Immunity in Bacteria and Archaea* (Springer-Verlag, Berlin, 2013).
- C. R. Fernández, Francis Mojica, the Spanish scientist who discovered CRISPR. https://www.labiotech.eu/interviews/francis-mojica-crispr -interview/ (April 8, 2019).
- E. S. Lander, The heroes of CRISPR. *Cell* **164**, 18–28 (2016). [This paper is rather controversial. See, for example: T. Vence, "Heroes of CRISPR" disputed, *Scientist Magazine,* January 19, 2016, https://www.the-scientist .com/news-opinion/heroes-of-crispr-disputed-34188; and M. Morange,

Why Eric Lander's controversial paper "The Heroes of CRISPR" is not solid historical research, *American Scientist*, February 17, 2016, https://www .americanscientist.org/blog/macroscope/why-eric-lander%E2%80 %99s-controversial-paper-%E2%80%9Cthe-heroes-of-crispr%E2%80 %9D-is-not-solid-historical.]

pp. 249–251 On CRISPR and the bacterial immune system:
- H. Ledford, Five big mysteries about CRISPR's origins. *Nature News* **541**, 280 (2017).
- R. Sorek, C. M. Lawrence, B. Wiedenheft, CRISPR-mediated adaptive immune systems in bacteria and archaea. *Annual Review of Biochemistry* **82**, 237–266 (2013).

pp. 250–251 On the mechanisms of Cas9's target DNA recognition and cleavage:
- G. Palermo, C. G. Ricci, J. A. McCammon, The invisible dance of CRISPR-Cas9. *Physics Today* **72**, 30–36 (2019).
- F. Jiang, J. A. Doudna, CRISPR–Cas9 structures and mechanisms. *Annual Review of Biophysics* **46**, 505–529 (2017). This article includes informative computer animations of Cas9's structural changes, available on You-Tube: https://www.youtube.com/watch?v=XAtZEIyzd7g, https://www .youtube.com/watch?v=Ya_Xoom7YAY.
- Nature Video, CRISPR: Gene editing and beyond, https://www.youtube .com/watch?v=4YKFw2KZA5o (2017).

p. 251 "Three papers published in 2005 . . . announced that the DNA sequences of spacers matched nonbacterial and nonarchaeal DNA sequences"
- F.J.M. Mojica, C. Díez-Villaseñor, J. García-Martínez, E. Soria, Intervening sequences of regularly spaced prokaryotic repeats derive from foreign genetic elements. *J. Mol. Evol.* **60**, 174–182 (2005).
- C. Pourcel, G. Salvignol, G. Vergnaud, CRISPR elements in *Yersinia pestis* acquire new repeats by preferential uptake of bacteriophage DNA, and provide additional tools for evolutionary studies. *Microbiology* **151**, 653–663 (2005).
- Bolotin, B. Quinquis, A. Sorokin, S. D. Ehrlich, Clustered regularly interspaced short palindrome repeats (CRISPRs) have spacers of extrachromosomal origin. *Microbiology* **151**, 2551–2561 (2005).

p. 252 "Reporting the results in a 2012 paper"
- M. Jinek, K. Chylinski, I. Fonfara, et al., A programmable dual-RNA–guided DNA endonuclease in adaptive bacterial immunity. *Science* **337**, 816–821 (2012).

p. 252 "awarded at a celebrity-laden Silicon Valley gala"
- C. Zimmer, CRISPR natural history in bacteria. *Quanta Magazine*, February 6, 2015. https://www.quantamagazine.org/crispr-natural-history -in-bacteria-20150206/.

p. 252 "another research group, led by Virginijus Šikšnys at Vilnius University, Lithuania, submitted a paper on April 6 also describing RNA-guided DNA cleavage"

- G. Gasiunas, R. Barrangou, P. Horvath, V. Šikšnys, Cas9–crRNA ribonucleoprotein complex mediates specific DNA cleavage for adaptive immunity in bacteria. *PNAS* **109**, E2579–E2586 (2012).

p. 253 "the forgotten man of CRISPR"

- S. Begley, Who gets credit for CRISPR? Prestigious award singles out three. STAT (website), May 31, 2018. https://www.statnews.com/2018/05/31/crispr-scientists-kavli-prize-nanoscience/.
- R. Bichell, Science rewards eureka moments, except when it doesn't. NPR .org, November 2, 2016. https://www.npr.org/sections/health-shots/2016/11/02/500331130/science-rewards-eureka-moments-except-when -it-doesnt.

p. 253 "he shared with Doudna and Charpentier the million-dollar Kavli Prize for Nanoscience"

- G. Guglielmi, Million-dollar Kavli Prize recognizes scientist scooped on CRISPR. *Nature* **558**, 17–18 (2018).

p. 253 "the labs of Feng Zhang at the Broad Institute of MIT and Harvard University and George Church at Harvard Medical School demonstrated CRISPR/Cas9 in mouse- and human-derived cells"

- L. Cong, F. A. Ran, D. Cox, et al., Multiplex genome engineering using CRISPR/Cas systems. *Science* **339**, 819–823 (2013).
- P. Mali, L. Yang, K. M. Esvelt, et al., RNA-guided human genome engineering via Cas9. *Science* **339**, 823–826 (2013).

pp. 254–255 On prime editing:

- V. Anzalone, P. B. Randolph, J. R. Davis, et al., Search-and-replace genome editing without double-strand breaks or donor DNA. *Nature* **576**, 149–157 (2019).
- H. Ledford, Super-precise new CRISPR tool could tackle a plethora of genetic diseases. *Nature* **574**, 464–465 (2019).

p. 256 On CRISPR and tomatoes:

- A. Zsögön, T. Čermák, E. R. Naves, et al., De novo domestication of wild tomato using genome editing. *Nature Biotechnology* **36**, 1211–1216 (2018).

p. 256 "researchers at the University of Pennsylvania piped blood cells and circulating immune cells"

- P. Tebas, D. Stein, W. W. Tang, et al., Gene editing of CCR5 in autologous CD4 T cells of persons infected with HIV. *New England Journal of Medicine* **370**, 901–910 (2014).

p. 257 "Another pioneering application in 2015 also involved immune cells . . . placed into a one-year-old girl who suffered from leukemia and was unresponsive to all other treatments"

- S. Reardon, Leukaemia success heralds wave of gene-editing therapies. *Nature News* **527**, 146 (2015).

p. 257 "In early 2020, a person with a rare genetic mutation that causes blindness became the first human to receive direct delivery of CRISPR/Cas9"

- H. Ledford, CRISPR treatment inserted directly into the body for first time. *Nature* **579**, 185–185 (2020).
- M. L. Maeder, M. Stefanidakis, C. J. Wilson, et al., Development of a gene-editing approach to restore vision loss in Leber congenital amaurosis type 10. *Nature Medicine* **25**, 229–233 (2019).

p. 258 "Anti-CRISPRs were discovered around 2012"

- J. Bondy-Denomy, A. Pawluk, K. L. Maxwell, A. R. Davidson, Bacteriophage genes that inactivate the CRISPR/Cas bacterial immune system. *Nature* **493**, 429–432 (2013).
- E. Dolgin, The kill-switch for CRISPR that could make gene-editing safer. *Nature* **577**, 308–310 (2020).

p. 258 "There are a multitude of different anti-CRISPRs"

- N. D. Marino, R. Pinilla-Redondo, B. Csörgő, J. Bondy-Denomy, Anti-CRISPR protein applications: Natural brakes for CRISPR-Cas technologies. *Nature Methods* **17**, 471–479 (2020).
- L. Liu, M. Yin, M. Wang, Y. Wang, Phage AcrIIA2 DNA mimicry: Structural basis of the CRISPR and anti-CRISPR arms race. *Molecular Cell* **73**, 611–620.e3 (2019).
- J. Shin, F. Jiang, J.-J. Liu, et al., Disabling Cas9 by an anti-CRISPR DNA mimic. *Science Advances* **3**, e1701620 (2017).

p. 259 "delivering an anti-CRISPR after Cas9 effectively shuts down further gene editing and hence reduces unwanted genetic alterations"

- M. Nakamura, P. Srinivasan, M. Chavez, et al., Anti-CRISPR-mediated control of gene editing and synthetic circuits in eukaryotic cells. *Nature Communications* **10**, 194 (2019).

CHAPTER 16. DESIGNING THE FUTURE

p. 260 "the horse and the sea-horse, the dog and the dog-fish, the snake and the eel"

- T. H. White, *The Book of Beasts: Being a Translation from a Latin Bestiary of the Twelfth Century* (UW-Madison Libraries Parallel Press, Madison, WI, 2002).

p. 260 "whole ecosystems are also subject to similar rules." See, for example:
- S. B. Carroll, *The Serengeti Rules* (Princeton University Press, Princeton, NJ, 2016).

p. 262 "Editing embryos was successfully demonstrated in mice and zebrafish in 2013"
- H. Wang, H. Yang, C. S. Shivalila, et al., One-step generation of mice carrying mutations in multiple genes by CRISPR/Cas-mediated genome engineering. *Cell* **153**, 910–918 (2013).
- W. Y. Hwang, Y. Fu, D. Reyon, et al., Efficient genome editing in zebrafish using a CRISPR-Cas system. *Nature Biotechnology* **31**, 227–229 (2013).

p. 262 "and in monkeys in 2014"
- Y. Niu, B. Shen, Y. Cui, et al., Generation of gene-modified cynomolgus monkey via Cas9/RNA-mediated gene targeting in one-cell embryos. *Cell* **156**, 836–843 (2014).

p. 262 "Chinese researcher He Jiankui stunned the world with the announcement that his team had used CRISPR/Cas9 to produce the first ever gene-edited babies."
- D. Cyranoski, The CRISPR-baby scandal: What's next for human gene-editing. *Nature* **566**, 440–442 (2019).
- Outrage intensifies over claims of gene-edited babies. NPR.org, December 7, 2018. https://www.npr.org/sections/health-shots/2018/12/07/673878474/outrage-intensifies-over-claims-of-gene-edited-babies.
- D. Normile, Chinese scientist who produced genetically altered babies sentenced to 3 years in jail. *Science*, December 31, 2019. https://www.sciencemag.org/news/2019/12/chinese-scientist-who-produced-genetically-altered-babies-sentenced-3-years-jail.

p. 262 "extremely abominable in nature"
- S. Jiang, H. Regan, J. Berlinger, China suspends scientists who claim to have produced gene-edited babies. CNN, November 29, 2018. https://www.cnn.com/2018/11/29/health/china-gene-editing-he-jiankui-intl/index.html.

p. 265 On baby carrots:
- R. A. Ferdman, Admit it, you didn't know this about baby carrots. *The Independent*, January 13, 2016. http://www.independent.co.uk/life-style/food-and-drink/news/admit-it-you-didn-t-know-this-about-baby-carrots-a6810651.html.

p. 265 On the Haber-Bosch process:
- J. W. Erisman, M. A. Sutton, J. Galloway, et al., How a century of ammonia synthesis changed the world. *Nature Geoscience* **1**, 636–639 (2008).

- S. K. Ritter, The Haber-Bosch reaction: An early chemical impact on sustainability. *Chemical & Engineering News* **86**, no. 33 (August 18, 2008). https://cen.acs.org/articles/86/i33/Haber-Bosch-Reaction-Early -Chemical.html.

p. 265 "About 40% of the earth's land is used for agriculture"
- J. Owen, Farming claims almost half earth's land, new maps show. *National Geographic*, December 8, 2005. https://www.nationalgeographic .com/news/2005/12/agriculture-food-crops-land/.

p. 266 On cane toads:
- *National Geographic*, Cane toad. https://www.nationalgeographic.com /animals/amphibians/c/cane-toad/ (September 10, 2010).
- T. Butler, Cane toads increasingly a problem in Australia. https://news .mongabay.com/2005/04/cane-toads-increasingly-a-problem-in -australia/ (April 17, 2005).

pp. 266–268 On gene drives, including applications to mosquitoes:
- National Academies of Sciences, Engineering, and Medicine, *Gene Drives on the Horizon: Advancing Science, Navigating Uncertainty, and Aligning Research with Public Values* (National Academies Press, Washington, DC, 2016). https://doi.org/10.17226/23405.
- J. Champer, A. Buchman, O. S. Akbari, Cheating evolution: Engineering gene drives to manipulate the fate of wild populations. *Nature Reviews Genetics* **17**, 146–159 (2016).
- N. Wedell, T.A.R. Price, A. K. Lindholm, Gene drive: Progress and prospects. *Proceedings of the Royal Society B: Biological Sciences* **286**, 20192709 (2019).
- M. Scudellari, Self-destructing mosquitoes and sterilized rodents: The promise of gene drives. *Nature* **571**, 160 (2019).
- K. Kyrou, A. M. Hammond, R. Galizi, N. Kranjc, A. Burt, A. K. Beaghton, T. Nolan, A. Crisanti, A CRISPR–Cas9 gene drive targeting doublesex causes complete population suppression in caged *Anopheles gambiae* mosquitoes. *Nature Biotechnology* **36**, 1062–1066 (2018).

p. 268 "Each year over 400,000 people die from malaria and more 200 million are infected"
- World Health Organization, World malaria report, 2018. https://www .who.int/malaria/publications/world-malaria-report-2018/report/en/.

p. 268 "Over a billion male mosquitoes with a genetic variation lethal to their offspring have been set free in Brazil, Malaysia, and the Cayman Islands over the past decade"
- S. Milius, Genetically modified mosquitoes have been OK'd for a first U.S. test flight. *Science News*, August 22, 2020. https://www.sciencenews .org/article/genetically-modified-mosquitoes-florida-test-release.

- L. Winter, 750 Million GM mosquitoes will be released in the Florida Keys. *Scientist Magazine*, August 21, 2020. https://www.the-scientist.com/news-opinion/750-million-gm-mosquitoes-will-be-released-in-the-florida-keys-67855.
- R. Lacroix, A. R. McKemey, N. Raduan, et al., Open field release of genetically engineered sterile male *Aedes aegypti* in Malaysia. *PLOS ONE* **7**, e42771 (2012).

p. 268 "Oxitec . . . notes that their deployment has been successful, causing, for example, an 80% decline in the mosquito population at its Grand Cayman test site"

- N. Gilbert, GM mosquitoes wipe out dengue fever in trial. *Nature*, November 11, 2010. http://blogs.nature.com/news/2010/11/gm_mosquitoes_wipe_out_dengue.html.

p. 268 "a 90% decline in Jacobina, Brazil"

- K. Servick, Study on DNA spread by genetically modified mosquitoes prompts backlash. *Science*, September 17, 2019. https://www.sciencemag.org/news/2019/09/study-dna-spread-genetically-modified-mosquitoes-prompts-backlash.

p. 268 "In 2021, the insects were released in the Florida Keys"

- E. Waltz, First genetically modified mosquitoes released in the United States. *Nature* **591**, 175–176 (2021).
- D. Coffey, First genetically modified mosquitoes released in U.S. are hatching now. *Scientific American*, May 14, 2021. https://www.scientificamerican.com/article/first-genetically-modified-mosquitoes-released-in-u-s-are-hatching-now/.

p. 269 "Oxitec's . . . lethal variation is not in a gene that encodes some biochemically specific activity, but in a transcription factor"

- K. Servick, Brazil will release billions of lab-grown mosquitoes to combat infectious disease. Will it work? *Science*, October 13, 2016. https://www.sciencemag.org/news/2016/10/brazil-will-release-billions-lab-grown-mosquitoes-combat-infectious-disease-will-it.

p. 270 "such as DNA-encoded mechanisms to block or even reverse gene drives in a population"

- M. R. Vella, C. E. Gunning, A. L. Lloyd, F. Gould, Evaluating strategies for reversing CRISPR-Cas9 gene drives. *Scientific Reports* **7**, 11038 (2017).

p. 270 "amateur enthusiasts . . . explore biotechnological crafts from their garages"

- H. Ledford, Garage biotech: Life hackers. *Nature* **467**, 650–652 (2010).

p. 270 On the Open Insulin Project:

- https://openinsulin.org/.

Index